D1200995

DISCARD

The Science of Love

By the same author

Grooming, Gossip and the Evolution of Language

The Trouble with Science

The Human Story

How Many Friends Does One Person Need?

The Science of Love

Robin Dunbar

WILEY

John Wiley & Sons, Inc.

Cover image: science beakers © Lumina Imaging/Getty Images; painting "The Kiss" (detail) by Gustav Klimt
Cover design: Wendy Mount

Published by John Wiley & Sons, Inc., Hoboken, New Jersey
First published in Great Britain in 2012 by Faber and Faber Limited

For general information about our other products and services, please contact our Customer Care Department within the United States at (800) 762-2974, outside the United States at (317) 572-3993 or fax (317) 572-4002.

Wiley also publishes its books in a variety of electronic formats and by print-on-demand. Some content that appears in standard print versions of this book may not be available in other formats. For more information about Wiley products, visit us at www.wiley.com.

Library of Congress Cataloging-in-Publication Data:

Dunbar, R. I. M. (Robin Ian MacDonald), date.
  The science of love / by Robin Dunbar.
    p. cm.
  Includes bibliographical references and index.
  ISBN 978-1-118-39765-7 (cloth); ISBN 978-1-118-46684-1 (ebk);
  ISBN 978-1-118-39767-1 (ebk); ISBN 978-1-118-39766-4 (ebk)
    1. Love. 2. Man-woman relationships. I. Title.
  BF575.L8D86 2012
  306.7—dc23

                                              2012025736

Printed in the United States of America

10 9 8 7 6 5 4 3 2 1

For Patsy

Contents

1

Now We Are One

O my luve's like a red, red rose
That's newly sprung in June!
O my luve's like the melodie
That's sweetly play'd in tune!

'A Red, Red Rose'

It's the weirdest thing that will ever happen to you. Falling in love, I mean. Think about it – there you are, wending your way innocently through childhood, doing the things that children do, and then the hormones suddenly kick in. And then you fall in love. Hesitatingly in that first all-consuming crush, but then with more confidence and determination as practice and experience make perfect. And although it doesn't happen every day, from time to time throughout the rest of your life it will catch you by surprise. It's very weird. All at once, you can't think of anything else except this seemingly random person who has just stepped – probably equally innocently – into your life. Your attention is focused almost to exclusion on the object of your desire. You just cannot get enough of them. You experience heightened happiness, often associated with glazed eyes, a faraway look and a dreamy expression, and roused (though not turbulent) emotions. The word 'besotted' often comes to mind.

Think Romeo and Juliet. Here in one single story, Shakespeare has managed to encapsulate every aspect of that extraordinary phenomenon in a beautifully crafted play. Were two star-crossed lovers ever so finely drawn?

Their agony and angst distilled so deftly? Their story remains the quintessential tale of unrequited love, of the unattainable for which the protagonists ache with such all-consuming passion. In this richly complex play, Shakespeare combines all the elements of the human mate choice predicament – the accidental meeting that precipitates instantaneous love on both sides, the friends that try to arrange trysts (as Benvolio does in his attempt to facilitate Romeo's meeting with Rosaline at the Capulet ball, thus inadvertently engineering the fatal meeting between hero and heroine), parents' inevitable attempts to manipulate their offspring's marital arrangements to their best advantage (as Capulet does in agreeing to Count Paris's request for his daughter Juliette's hand), and, last but not least, the raw uncertainty as to whether we can achieve our desired outcome (instantiated here by the enmity that separates the Capulet and Montague families and forms so insidious a barrier between the lovers).

The story raises, in one seminal moment, all the questions about love and betrayal that one can imagine. Why did Romeo fall so suddenly for Juliet, when he went to the ball to sneak a meeting with Rosaline? Can we really fall in love at first sight, or is that just an urban myth? Why is the desire for a kiss so strong? Can we really love one person forever? Are we ever so distraught that we could take our own lives when our passion is thwarted – never mind if we return home to discover the love of our life awaiting burial? But even if we don't go to the lengths that Romeo did, can we actually 'die of a broken heart'? And even if we can't, why is it that we feel the pain of separation or rejection as real pain?

This book brings modern science to bear on these questions. It will oblige us to draw on scientific disciplines that are very rarely bedfellows. The very richness of the experience makes that inevitable. But first, what is this phenomenon we refer to as falling in love? And is it really a human universal? Many have claimed that it is not, and that many traditional societies do not recognise it – that it is a phenomenon born of nineteenth-century romantic novels. So let's begin by looking more closely at what we have to study.

The kind of besottedness that we associate with romantic love can be both intense and, compared to mate attraction in most other animals, relatively long-lasting. This early intense phase of a human relationship typically lasts twelve to eighteen months, but can often extend for several years beyond that in attenuated form. In the heady intellectualised aftermath of the 1960s, it became fashionable among intellectuals, and especially among anthropologists, to assert that this sense of falling in love is a peculiarity of modern, Western, capitalist culture, driven no doubt by the market in Mills & Boon-style romantic fiction. In traditional societies, people did not marry for love, but as a matter of economic convenience or for political reasons. It is still a common view. But this is to confuse the reasons for marriage contracts with the relationships involved. People have always been hard-nosed and married for political or economic convenience. Arranged marriages have been a feature of every human culture the world over. Currently, they happen to be especially common throughout much of South Asia, from Iraq as far east as Japan, but they were the bread and meat of the noble houses of Europe ever since the Romans left us alone to get on with our lives as best we could. People still marry for convenience and economic advantage every day all over the Western world. But that doesn't mean to say that people don't fall in love. Whether they *marry* because they fall in love is a separate issue. In actual fact, the falling in love bit can happen just as often the other way around: people get married for strategic reasons and then, lo and behold, fall in love. As Molière put it in his play *Sganarelle* (1660): 'Love is often a fruit of marriage.'

The evidence from arranged marriages tells us that the seemingly hapless couple often end up falling in love with each other after the formalities of the wedding – sometimes months, sometimes even years later. Arranged marriages are no more likely to be soulless forms of socially sanctioned prostitution than those in which the happy couple thought they had married for love. Many, if not most, of those in arranged marriages fall in love with the partner they are saddled with. We in the post-Romantic West assume we have choice over whom we fall in love with and marry. But, in reality, our choice, as I shall show in the chapters that follow, is actually somewhat random and decidedly limited – after all, we rarely search through more than a handful of potential spouses before finally choosing 'the one'. It's really just a question of when you do the falling in love bit – before or after agreeing to marry the person. Yet even in the supposedly liberated West, not all of us have this experience. Many of us make do with whatever we can get . . . and grit our teeth. But that doesn't mean that the phenomenon of falling in love is a social construction that people only experience because they've been told they ought to.

The truth is that, notwithstanding vigorous claims to the contrary, some form of romantic attachment transcends historical and cultural boundaries, and may well be a human universal (accepting that there are degrees of expression in this trait even within the same culture).

> Tossed and bewildered, like a flickering candle,
> I roam about in the fire of love;
> Sleepless eyes, restless body,
> neither comes she, nor any message.
> In honour of the day I meet my beloved
> who has lured me so long, O Khusro.

wrote the medieval Indian poet Amir Khusrau Dehlavi (1253–1325). At around the same time – and long before the era of Mills & Boon – we find the celebrated French troubadour Guillaume de Marchaut (1300–77):

> For I love you so much, truly,
> That one could sooner dry up
> the deep sea
> and hold back its waves
> than I could contain myself
> from loving you . . .

And in another of his songs:

> Sweet noble heart, pretty lady,
> I am wounded by love
> So that I am sad and pensive,
> And have no joy or mirth,
> For to you, my sweet companion,
> I have thus given my heart.

Or from the Sanskrit of the fifth-century Indian poet Kālidāsa comes this evocative quatrain:

> Sloe-eyed, please stop for a moment
> Tying up prettily those locks of hair;
> For my eyes are entangled there,
> I have been extricating them the whole day.

Or earlier still, from around 900 BC, is the author of the Bible's 'Song of Songs' (or, as it is sometimes known, 'Song of Solomon'), who had this to say:

> O that you would kiss me
> with the kisses of your mouth!
> For your love is better than wine,
> your anointing oils are fragrant,
> your name is oil poured out.

And later in the same series of poems (for that is clearly what they actually are):

> How graceful are your feet in sandals,
> O queenly maiden!
> Your rounded thighs are like jewels,
> The work of a master hand.
> Your navel is a rounded bowl
> That never lacks mixed wine.
> Your belly is a heap of wheat,
> Encircled with lilies.
> Your two breasts are like two fawns,
> Twins of a gazelle.
> Your neck is like an ivory tower.

Your eyes are pools in Heshbon,
By the gate of Bath-rab'bim.

. . . and so it goes, on and on.

Or, even earlier, there is the Egyptian pharaoh Rameses the Great, who more than 3,200 years ago wrote this on the tomb of his favourite wife, the politically powerful Nefertari (not to be confused with the even more famous Nefertiti, who lived about a century earlier): 'The one for whom the sun shines . . .' How often has *that* been said in history, and not by folks who just happened mysteriously to know how to read Egyptian hieroglyphics deep underground in a queen's burial chamber centuries before they were eventually deciphered? And here, from around 2025 BC, translated from the cuneiform script on a modest-sized tablet discovered in 1889 during excavations at the Sumerian city of Nippur in modern-day Iraq, is what is often claimed to be the oldest love poem in the world:

Bridegroom, let me caress you,
My precious caress is more savoury than honey,
In the bedchamber, honey-filled,
Let me enjoy your goodly beauty,
Lion, let me caress you,
My precious caress is more savoury than honey.

According to Samuel Krame in his book on the Sumerians, that sense of love was by no means uncommon, despite the fact that marriages in Sumer were invariably arranged on economic grounds – literally measured in silver.

The truth is that falling in love *is* a human universal: it occurs in every culture and every time, and has done throughout eons of human history back to that distant moment when some ancient Eve awoke one morning . . . and melted at the sight of the Adam before her. That doesn't mean that all of us experience it, or even that all of us have these experiences under circumstances that eventually lead to marriage – or whatever the appropriate social equivalent might be. But it does mean that it happens often and frequently. And it does seem to be important. Sandra Murray and her colleagues have been studying romantic relationships now for several decades, and have found that idealising one's partner is a sure recipe for marital success; moreover, the higher one's ideals are and the more one idealises one's partner, the more satisfied one is with the relationship – and the longer it is likely to last. And this isn't because people who idealise their partners more have more deserving partners: in fact, there appears to be only a modest correlation between an individual's perception of their partner and the partner's evaluation of their own traits. It seems as though there is something about the intensity of this peculiar phenomenon that *is* important to the success of the whole endeavour. And that creates a puzzle: if it's so universal a characteristic of humans, it must have a biological basis and a biological function. Yet it has been effectively ignored by most scientists. We don't really know what it is, or why we have it – or even whether it bears any resemblance at all to the kinds of things that other animals experience.

It's that biological story that will be the subject of this book. Part of that story will lie in understanding what it is that causes us to feel like this. But I'm not just going to be interested in the obscure aspects of physiology or genetics that underpin this curious phenomenon. There is more to biology than that. My interest will lie in trying to bridge the gap between these more obviously biological bases of our behaviour and the psychological, social, historical and evolutionary contexts that help to mould that behaviour. Not least among these are the principles that underlie our choice of mates, and the tactics we use to ensnare them once we have spotted them. We'll take a very broad look at the business of falling in love, and ask the poets for a bit of help along the way. Mostly, but not necessarily of course, such relationships are heterosexual, but my guess is that the underlying processes that produce these relationships are not that different between the two sexualities (and all combinations between the two extremes), so I shall simply take it for granted and have no more to say about this particular issue.

Can't take my eyes off of you ...

Everyone appreciates that when we fall in love our attention seems to become focused on one person to the exclusion of all others. But there has been an ongoing debate as to whether this is due to the fact that falling in love causes you to lose interest in other members of the opposite sex (the deflection hypothesis) or to you just becoming so wrapped up in your new love that you don't have time to attend to anyone else (the attention hypothesis). The difference may seem slight, but it belies an important contrast in the underlying psychology. Under the deflection hypothesis, you lose any motivation to be interested in someone else, whereas under the attention hypothesis some accident of circumstance might lead you to notice someone else and so switch your attention to them and abandon your previous mate. In effect, the first implies a psychological mechanism that actively inhibits your likelihood of being attracted to someone else, whereas the second does not.

In the normal course of things, our attention is remarkably easily distracted by members of the opposite sex. We naturally check them out more often than we do members of our own sex. Some years ago, I and my students undertook a series of studies on social monitoring behaviour in the refectory of a large London university and in nearby parks and public gardens. We were interested in testing between four hypotheses that might explain why people occasionally glance around their immediate environment, even when busy eating or engaged in a conversation. The four hypotheses were: checking for predators (in this context, people who presumably might attack or rob you), checking to see whether any friends had turned up, checking for potential new mates, and checking for rivals who might steal a mate away. The third hypothesis – checking for members of the opposite sex who might offer opportunities of new romantic relationships – won hands down. The patterns of when and how people glanced around them, and whom they looked at when doing so, uncompromisingly pointed the finger at mate choice as the explanation. We found that men and women were much more likely to look up and glance at someone moving near them if that individual was of the opposite sex than if they were of the same sex. In fact, they were so sensitive to the sex of the individual concerned that they were often aware of the sex of someone coming behind them long before they came into full view. Our peripheral vision, it seems, is extraordinarily good at picking up subtle cues of gender.*

---

* For present purposes, I will use the terms *sex* and *gender* as meaning much the same thing, following biological practice.

When people are in love, they spend much less time looking at attractive members of the opposite sex, and they rate those whom they do see as less attractive than do people who are currently single. In a rather ingenious study, Jon Maner and his colleagues showed subjects photographs of attractive and average same- and opposite-sex individuals in one corner of a computer screen, and then asked them to perform a task that required them to switch attention to another part of the screen. Those who were in love were much faster at switching attention away from opposite-sex photos of attractive individuals than people who were single, even though there was no difference when attending to photos of averagely attractive members of the opposite sex or photos of anyone of the same sex. They concluded that when in a romantic relationship, our attention seems to be actively repelled from serious rivals. It seems that we actively downgrade people we would normally consider physically attractive to the same level of attractiveness as the jobbing average.

Johann Lundström and Marilyn Jones-Gotman used odour to gain more insights into this issue. They asked young women who had a romantic partner to complete a questionnaire about the depth of their love for that person and then asked them to try to distinguish between the odours of their boyfriend, another friend of the opposite sex and a female friend. The odours had been collected by asking the various individuals concerned to sleep (alone) in a cotton T-shirt with nursing pads under the armpits to absorb natural odour for seven nights in a row. There was no correlation between the intensity of the women's romantic involvement with their boyfriend and their ability to identify either their boyfriend's odour or the female friend's odour, but there was a significant *negative* relationship with their ability to identify the odour of the opposite-sex friend. In other words, being romantically involved actually seems to turn you off from potential rivals rather than causing you to be *so* besotted with the object of your love that you just forget to be interested in anyone else.

Motherhood and apple pie

Romantic relationships aren't the only intense relationships we have, of course. The most obvious of these must be the incredibly strong bond that develops between a mother and her offspring. Indeed, we speak of 'mother love', using the same term that we use for romantic relationships as though it were much the same kind of thing. Here, surely, is another human universal, though one that, as with falling in love, varies a lot in its intensity from one individual to another. There is no question, for example, about the fact that men just don't have quite the same intensity of feeling that women have about small children, especially babies, even when they are their own. Women – or at least some women – are just more maternal than men, even if it is also true that some men are more maternal than other men, and perhaps even than some women. Nonetheless, these individual differences should not blind us to the fact that here is another deeply intense emotional effect that binds us irrevocably to another individual, that this is a human universal (indeed, it may even be true of most mammals) and that it is there for a reason.

Mother-and-baby relationships share many of the features of romantic relationships: the completely focused attention; the sense of wonderment, elation and contentedness;

the wanting just to be with the object of desire, to touch and stroke the person concerned; the willingness to do anything for them. This might lead us to suppose that the processes involved in romantic relationships had their origins in mother–infant bonds. The same chemistry has just been generalised from one context to another. This is not an implausible suggestion. After all, you might suppose, the business of making babies and the romantic feelings that underpin this are just the prelude for the bigger job of getting that baby through to independence, so why not just use the same psychological machinery and adapt it to the first part of the process?

The maternal instinct is fundamental for mammals, whose key evolutionary adaptation was the two-stage process of internal gestation and then lactation. Internal gestation is not entirely unknown elsewhere in the animal kingdom. Famously, male sea horses bear and give birth to live young; cichlid fish suck in their eggs and allow their young to develop within their mouths, while midwife toads insert their eggs under their skin and allow the tadpoles to develop there in little blisters. Nonetheless, internal gestation is relatively rare and rather patchy in its distribution across the animal kingdom. It is only in mammals that all species are obligate internal breeders.* One of the benefits of this rather challenging option is the ability to produce large-brained offspring who are then capable, as adults, of fine-tuning their behaviour more effectively to the particular circumstances in which they later find themselves. This early nurturing is critical because

* The only exceptions, of course, are the monotremes, the group of rather primitive egg-laying mammals that includes the platypus and the echidna.

it allows the parent to invest in brain and body growth over a much longer period of time. Since brain tissue can only be laid down at a constant rate, the key to evolving a big brain lies in extending the period of parental investment for long enough to get the brain you ultimately want.

This whole business is very taxing and demanding for the parents who have to keep up this effort rather longer than they would really like to. This isn't, perhaps, so much of a problem during gestation, when the baby is small and inside one's body. But once it has been born, its energy demands rise exponentially as it grows rapidly to the size where it can safely fend for itself. However, in some species like monkeys and apes there is a whole additional phase of socialisation that kicks in after the end of the lactation period. Shepherding one's young through the business of acquiring the requisite social skills and placing them as judiciously as possible in the adult social world only begins when the baby is weaned and it can carry on for many, many years thereafter. That all adds up to what has to become a labour of love – especially during the early months when babies don't themselves provide that reinforcement. This is a particular problem for humans because our babies are born so prematurely that it takes them about a year to reach the same stage of development and independence as other monkeys and apes at birth. Monkey and ape babies can get up and stagger about within days, if not hours of birth, but our babies are so neurally immature that most cannot do this much before their first birthday.

This is a curious evolutionary side effect of the fact that our ancestors first adopted an upright stance. This caused a remodelling of our pelvis to provide a stable platform to support the trunk, and a consequent narrowing of the birth canal. Then, several million years later, our brains began to undergo a dramatic increase in size. The result was a square-peg-in-a-round-hole kind of problem: any significant increase in the baby's brain size at birth meant the head was much too big to get through the now rather modest-sized birth canal. This mightn't have been such a problem, but an accident of history hundreds of millions of years earlier when our fishy and reptilian ancestors were first evolving had resulted in the reproductive tract passing through what became the bones of the pelvis rather than over them. A more sensible arrangement would have been to have the urethra and reproductive tract coming out just below our belly button. That would have saved no end of problems later. But evolution is not that good on foresight, and later generations are often stuck with the unfortunate consequences of past evolutionary events.

Our solution was to give birth to as premature an infant as we could get away with. By monkey and ape standards, our infants are desperately, even dangerously, premature when they are born and they can survive that crucial first year outside the womb only by dint of some very devoted parenting – and, in particular, by a mother who is, in the normal state of nature, willing to carry on lactating with seemingly enthusiastic abandon until the baby is old enough to start feeding for itself. Something pretty powerful is needed to get us over that hump and persuade us to keep pumping milk and food into the ever-open maw.

In fact, mother–infant relationships aren't the only kind of interaction we have that seems to share something with romantic attachments. Though the meaning of the word has been rather debased of late by Facebook, friendships are a third category of close relationship that we have.

Intimate friendships, in particular, share many of the deeper and more meaningful features of romantic relationships, so much so that occasionally they can even spill over into full-blown sexual relations. We tend to distinguish different kinds of friends – intimates, good friends, down to the 'he's just a friend' type – and for the good reason that this is in fact a graded series of relationships. There is good evidence that we find it difficult to maintain more than one genuinely committed romantic relationship at a time. But we can have many kinds of friendships, which grade imperceptibly into each other.

A third important category is kinship. And I don't just mean the feelings of love for one's parents that eventually allow us to reciprocate for those devoted years of parental solicitude. Kinship is weird stuff. For most of us, something genuinely visceral kicks in when we discover that someone is related to us, however distantly that might be. It suddenly puts a perfectly ordinary stranger in a completely different class. They are no longer just 'people', but kin, those with whom you share blood. Just on the strength of that single ephemeral three-letter word you would have them round to dinner, take them in, or even lend them your car. And I don't mean just rediscovering your birth parents decades after being adopted. I mean distant kin like third cousins, people with whom you share just a great-great-grandparent. What is even weirder about this is that the relationship is a purely linguistic one, something we can only specify in language. There is nothing really tangible about it, because neither you nor they knew your common great-great-grandparents – they are folk you've simply heard about by word of mouth. Almost none of us have ever had the chance to meet a great-great-grandparent, to be dandled on their knee, or spend time with them. Yet when the magic word trips off the tongue, suddenly you are blood brothers, bound by a common bond.

We are much more embedded in kinship circles than we often realise. Our research suggests that around half the people we consider dear to us are in fact members of the family, by blood or marriage. And by this, I don't mean just your mum and dad and siblings, but those in your extended family right out to second cousins and beyond. For the average person, that amounts to around fifty to seventy people in all. We give them priority above all others, are willing to lay our lives down for them if push comes to shove in a way we simply wouldn't even consider for people who were not related to us. The Inuit (or Eskimo) of the Alaska coast still go whaling in small open boats, Moby Dick style, as they have done for centuries. It is an exceptionally dangerous business, since the boats often get upended by whales as the whalers get in close enough to be able to harpoon the leviathans by hand. Because the risks of being thrown into the water are so high, the crews always consist of close kin – because, say the Inuit, no one but a close relative is willing to dive into icy Arctic waters to rescue you if you get thrown out of the boat. Blood *is* thicker than water.

Just how does a romantic relationship differ from a friendship, or a friendship from kinship? And how are close friends different from friends-of-friends, or close kin different from distant kin? I have spent the last decade exploring the structure of our social world, and this book will draw heavily on my work to tease out a framework for understanding the similarities and differences between our various kinds of relationships. At its core will lie the role of trust in allowing two individuals to form a bond that provides mutual support as well as pleasure. But, first, there is one more issue to put on the table.

What is love?

Freud, inevitably, had a great deal to say about love, and there has been a long (and, on the whole I think, largely unhelpful) tradition of clinical interest in the topic. In this broad approach there has been a general distinction between *eros* and *agape*, terms borrowed from classical Greek to refer to erotic, sensual love and something closer to friendship with its connotations of altruism and unstinting generosity (and from this adapted in the Christian tradition to refer to religiously motivated love, usually towards the Redeemer or one of his saints). Most of this psychoanalytical literature is pretty dull, and has in any case tended to focus on the negative aspects of inadequate or poor-quality relationships, so I will ignore it in favour of the other group of people who have been interested in romantic relationships – and to a lesser extent, friendships – namely, social psychologists.

There have probably been two main approaches among social psychologists. One, attachment theory, has a developmental focus and argues that our adult romantic and other intimate relationships develop out of, or are scaffolded by, our early experience of mother–infant relationships. Attachment theory largely derives from the influential British psychiatrist John Bowlby, who tried to combine psychiatric theories, including Freudian psychoanalysis, with ethological observation. He was responsible after the Second World War for persuading maternity hospitals to abandon the practice of placing newborn babies in nurseries away from their mothers from the moment they were born. He argued passionately that mother and baby needed to develop a close bond during the first few hours after birth and that it was absolutely essential for the future development and wellbeing of the baby that this was allowed to take place naturally, by leaving the baby with the mother. That's one reason why babies are now put straight to the mother's breast as soon as they are born, even before they have been cleaned up. Bowlby's ideas were extremely influential, and almost all of you reading this now have been the beneficiaries of his reforming crusade. But are your adult romantic relationships at root just your relationship with your mother writ large? The answer is, for perhaps obvious reasons, largely no – though, as I shall show later, your relationship with your parents does have unexpected implications for your choice of romantic partner later in life. But the bottom line is that neuroimaging suggests that maternal love and romantic love are actually two different things: they involve some of the same bits of the brain, but, importantly, they also involve some very different bits of the brain.

That said, however, it is a robust finding that the quality of a romantic relationship has considerable repercussions for one's emotional and psychological wellbeing, both in the teenage years and through adulthood. A common finding is that being in a relationship, and especially a congenial relationship, has a positive impact on self-esteem and psychological wellbeing, and that in turn almost certainly has a very significant impact on physical health and the body's ability to resist disease. Poor relationships and relationship breakdown, on the other hand, are both risk factors for depression. Similarly, the experience of relationships, both parent–child and sibling–sibling, within the home environment during childhood provides a framework that establishes the general pattern of our relationships as adults. A series of short-lived, unstable relationships in the teenage years is predictive of poor relationship quality in adulthood. It seems that failure to practise the skills needed for mature relationships is important: one of the best, and most robust, predictors of marital dissatisfaction and divorce is a teenage relationship that resulted in early marriage. That's not to say that all such relationships end in failure, but rather that statistically speaking it's not a good start.

The other approach within social psychology has focused almost entirely on how we see ourselves in relationships, with a particular interest in the emotional and cognitive components of a relationship and whether these predict relationship satisfaction, stability and longevity. Building on the very successful paradigm used to study, first, intelligence and, later, personality, the standard approach has been to ask people to agree or disagree with statements of the kind: 'I feel an emotional uplift when I see [person X]', or 'I experience great happiness when I am with [X]' or 'I can't imagine ever ending my relationship with [X]'. Answers to several hundreds of these kinds of questions by many thousands of people are then subjected to some heavy statistical analysis to look for consistencies and patterns. This identifies common themes in the answers that are interpreted as factors or dimensions. In personality theory, these constitute the familiar 'extraversion/introversion', 'neuroticism', 'openness', 'conscientiousness' and 'agreeableness' personality dimensions, commonly known as the 'Big Five'. While personality theory works reasonably well and personality profiles measured in this way are generally fairly consistent across time and space, social psychologists' attempts to disentangle the nature of love, romantic or otherwise, have been much less successful. As many of them have commented, few psychological constructs have been more elusive than the construct of love. Somehow, it has proved exceptionally difficult to put a finger on exactly what we mean by it, and so it has been all but impossible to describe it.

Perhaps the best and most successful of the attempts to define romantic relationships in this way is Robert Sternberg's 'Triangular Theory of Love'. He argued that romantic relationships can be categorised along three independent dimensions: intimacy, passion and commitment. Passion reflects those aspects of a romantic relationship commonly associated with what we think of as 'falling in love' (an intense focus on the object of one's desire and a sense of exhilaration, with or without a sexual component). Intimacy reflects feelings of closeness, connectedness and bondedness, while commitment reflects a desire to support, and remain in the physical presence of, another individual. The latter two have sometimes been referred to by the slightly more evocative terms 'feeling close' and 'being close'.

Sternberg's categorisation has the merit of allowing us to see how different relationships might vary on his three dimensions (see diagram on next page). Infatuated love, for example, is when passion is high, but intimacy and commitment are low, whereas romantic love derives from a combination of high intimacy and high passion in the absence of commitment. Companionate love is the combination of high intimacy and high commitment in the absence of passion, whereas fatuous love is the combination of passion and commitment in the absence of

The Science of Love

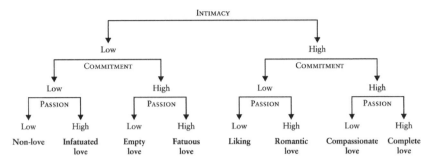

Sternberg's 'Triangular Theory of Love'. An extensive analysis of people's answers to a battery of questions about their relationships led Sternberg to conclude that relationships have three dimensions: intimacy, commitment and passion. Since each can be low or high, this gives rise to eight distinct types of relationship (shown in bold at the bottom of the diagram). These vary in quality from 'non-love' (disinterest or the absence of love), through 'liking' to various forms of 'infatuated', 'empty' and 'fatuous' love to 'compassionate' love, and finally to what he considered to be true, 'complete' (or consummated) love, where all three dimensions are high.

intimacy. When all three dimensions are high we get consummate or complete love. These make intuitive sense, though one might argue about the terms he uses for the different kinds of love. Nonetheless, they serve to remind us that relationships can come in many and varied forms and aren't necessarily all of a kind. Or at least, the motivations and emotional drivers that underpin them can vary considerably and give rise to relationships of different strengths and value.

This categorisation, imperfect as it is, is also a useful reminder that reciprocity is an important feature of all intimate relationships. Unrequited love is what bothers the poets most, it seems, if only because it arouses the most intense emotions – emotions of loss and unfulfilled desire. And some relationships can be quite one-dimensional, as in those based solely on lust (or, as Sternberg would have it in more polite vein, passion). But some of these may simply represent stages in the development of a relationship, from the initial unrequited love through consummated passion to friendship in more mature romantic relationships. However, useful as it is, it doesn't really get us beneath the surface of what a relationship actually is – what it is that produces all that turmoil of emotions, the abreaction when our tentative advances are spurned or rejected or a lover betrays us. Nor does it get us beyond the immediacy of our 'raw feels' to ask how we choose our lovers or our friends. And it certainly doesn't ask why these things exist in the first place. Like most of psychology, it takes the world as we find it as given, almost as though it were the Panglossian best of all possible worlds, and never asks about how or why it came to be this way. Asking the evolutionary questions about history and the biological functions that phenomena are designed to serve can often uncover unexpected oddities in the world that don't make a great deal of sense. Very often, these are – like our helplessly premature babies – evolutionary compromises that only make sense when seen in terms of the bigger picture.

When language is so inadequate

At the end of his book *The Symbolic Species*, the neuro-anatomist Terry Deacon observed that humans have very unusual living arrangements. They live in monogamous pairbonds that are set within a large social community in which many males and females live together. This would be fine, but for what has become known among anthropologists as the 'division of labour': from time to time, men and women go their separate ways, especially in traditional societies where males go hunting and females gather vegetable foods. The problem, as Deacon saw it, is that whenever a mated pair are apart, they are at risk of rivals who might either steal the mate or effect extra-pair copulations. This is a particular problem for males. Males are always vulnerable to paternity uncertainty: among mammals, a female always knows that the offspring she gives birth to are hers, but a male can never be 100 per cent certain. Deacon argued that humans face this problem in a particularly intrusive way. We are always surrounded by many rivals for the attentions of our romantic partner, and they have open season when the division of labour obliges one or other sex to leave their mates for long periods of time (e.g. while away hunting).

The solution, Deacon suggested, was overt social statements of ownership such as marriage ceremonies and symbolic badges. To signal marital status, we use a whole raft of purely symbolic markers, not least among which is the wearing of wedding rings. In many cultures, women also adopt a whole host of additional practices such as using titles like 'Mrs', adopting the husband's surname, and changes in style of dress or coiffure. In traditional Polynesia, a couple marked their marriage by placing *leis* (traditional flower necklaces) around each other's necks, and women switched from wearing a flower behind the right ear (meaning *Still available*) to wearing it behind the left (meaning *Spoken for*). Deacon argued that, being symbolic, these all require language, and he thus saw symbolic contracts of this kind as being the key selection pressure behind the evolution of language – hence the title of his book. If monogamy evolved early (and he assumed that it must have evolved very early), then language necessarily evolved early too – and by early, he meant with the appearance of *Homo erectus*, the first member of our genus, around two million years ago.

Deacon is surely right to identify the formation of monogamous pairbonds in humans as a major anomaly that requires explanation. But the real issue, and the one that lies at the focus of our enquiry, is why on earth pairbonds evolved at all, not how we keep them together in the face of the risks that threaten to engulf them. The conventional explanation has always been that it requires two to raise human offspring, and the assumption always was that two meant mum and dad. It is not entirely obvious to me that the costs of rearing inevitably require that it be mum and dad, though the frequency of pairbonded monogamy among the birds for precisely this reason might predispose us to thinking this is the obvious first call. An equally plausible possibility in the case of humans, at least, is mum and granny. After all, why do human females give up all possibility of reproducing just when they are at their peak of maternal experience and skill at around forty-five years of age? The menopause is all but unique to humans. There have been claims that chimpanzees and elephants – two other long-lived species – also undergo menopause, but to be honest this is more like a slowing down of reproduction in very old age.

Humans are clearly in a different league: no one else so completely gives up the possibility of reproduction so early in the lifespan. And the conventional explanation for that has always rested on the observation that the timing coincides with the point at which daughters start to repro-duce. Mum gives up reproducing herself so as to be able to help her daughters – the so-called grandmother hypothesis. As the same time, it provides a neat explanation for why mother–daughter bonds should be so important and should persist into adulthood – and why grannies should be so obsessed with their grandchildren. But, while this may explain one aspect of human behaviour, it leaves us with two puzzles. One is why pairbonds should form between a man and a woman, given that they appear to be unnecessary. The other is whether language and symbolic cognition are really necessary to create a system in which pairbonds exist within large communities that contain many individuals of both sexes. Is a relationship really a conscious, cognitive phenomenon that we can think deeply about and need language to regulate, or is it buried so deeply in emotion that it is almost opaque to our conscious mind?

The second issue is much the easiest to answer, because it turns out that the kinds of mating/social systems that Deacon worried about are not as rare as he imagined. In hamadryas baboons, for example, males form harems of up to four females, whose loyalty they enforce by punishment: females that stray too close to other males are given severe bites on the back of the neck that have them cowering close to their male for hours afterwards. Nonetheless, the integrity of the pairbonds between a female and her male are not just dependent on the harem male's proactive reprimands: much also depends on other males' willingness to test the strength of these bonds. Males normally 'respect' the pairbonds of other males and will assiduously avoid trying to mess with another male's female. A male placed in a cage with another male and female that he has watched interacting will sit at the edge of the cage showing intense interest in whatever might be going on outside the cage, or fiddling with the grass at his feet – *anything* but look at the other male's girl, since that would constitute an explicit threat and instantly precipitate a fight. This phenomenon (known as triadic differentiation) seems to be designed to protect the pairbond.

However, the extent to which rival males behave in this way depends on their assessment of the female's attachment to her male, something they pick up on by noticing how attentive the female is to her bonded male. If she shows limited interest in monitoring where he is, and doesn't seem to be paying him all that much attention, a rival male introduced into the cage will sometimes have a go at separating her from her male, if necessary by openly fighting for her with him. If the female is not that attentive to her male, then she may be willing to switch her interest to him – and that may be just enough to tip the balance against her previous male even if he is more dominant than the rival. In contrast, a male who senses that the female is constantly glancing at her male, and following him every time he moves, won't bother even to try. If she is strongly committed to her male, there will be no chance of wresting her from him even if the rival is physically stronger than her male. It is under these circumstances that males suddenly become so interested in events going on outside the cage or in the minutiae of their toes.

In a more explicitly monogamous species like the South American titi monkey, both sexes actively preserve the pairbond by discouraging rivals of the same sex from coming too close. Similar behaviour is seen in antelope like the klipspringer. This small African antelope, not a lot bigger than a weaned lamb, is intensely monogamous, perhaps one of the most intensely monogamous of all mammal species. Each pair lives in a tiny territory barely twice the size of a football pitch on a rocky outcrop in the savannah grasslands of eastern and southern Africa. Again, each sex is profoundly jealous of its partner, chasing off any stranger of the same sex that happens to drift onto its territory.

But there is one species that seems to have an almost identical social arrangement to humans, and this is the little bee eater, a tiny member of the bee eater bird family that builds nesting burrows in sandy banks along river courses in the savannah plains of East Africa. Because suitable nesting areas are few and far between, bee eaters are forced to congregate at the handful of places where they can successfully burrow into the earth. This usually means many hundreds, even thousands, of birds sharing the same bank. Even though each breeding pair has its own burrow, the burrows are densely packed and the female, in particular, has to run the gauntlet of unattached males hanging around the edge of the colony whenever she goes out to the foraging grounds to feed. To reduce the risk of harassment, the bee eaters have very close pairbonds so that the male accompanies his female wherever she goes. In effect, he acts as her bodyguard. And all this is done without a shred of language to manage it.

In so far as the purpose of these kinds of behaviour is to protect the pairbond (at least for a limited period of time), it should give us pause to ask why humans would need anything as complex as social contracts or language to do the same job. Instead, the fact that monkeys and birds can solve the same problem by purely behavioural means suggests that symbolic markers and language may not in themselves be essential for protecting pairbonds in social environments. What it might suggest, instead, is that pairbonds (and the processes that underpin them) evolved well before language, and that language was later co-opted to the business of reinforcing whatever natural mechanisms already existed to protect pairbonds.

I mention language because it brings me back to my starting point, and the poets whose evocative words we so admire. Most of us seem to have great difficulty in expressing our emotions in language. Words literally fail us just at the moment when we really need them. How often do we use the phrase 'You know what I mean?' in exasperation at our own inability to turn feelings and vague thoughts into words? Yet some people have the gift of being able to say just what we wanted to say, to put into words what we instantly recognise as the very feelings with which we struggled so inarticulately.

There are two important lessons here that are germane to our enquiry. One is that the emotions that well up and create our inner feelings are not well connected to the conscious, language-accessible brain. They belong to the emotional right side of the brain, the side that seems to handle our more supposedly irrational, animalistic reactions. Conventional neurobiological wisdom has it that our language capacity is, by and large, lodged in the left side of our brain, and it seems that its connections to those emotional centres in the right half aren't as good as they might be. This ought, I think, to alert us to the fact that all this falling in love stuff might arise from a bit of deeply buried emotional machinery that we don't acquire just by reading Mills & Boon novels. Rather, it is something that is very ancient, something we inherited from our remote ancestors long before they acquired language. The other lesson is that it offers us an explanation as to why we should revere poets. These rare individuals – and I think we can all accept that the ability to write good poetry *is* rare – seem to have the knack of accessing their emotional brain with their conscious mind and turning what they find there into language.

It is a rare and exceptional skill, and we rightly do them homage. But it reinforces the fact that we are just not very good at explaining what's going on inside us. We *feel* our emotions, but we do not always understand them. The problem for our present enquiry is that this makes it very difficult for us to dig beneath the surface and find out what is actually going on. It's a problem that has bedevilled all attempts to study romantic relationships – and, indeed, all other kinds of relationships – scientifically. Nonetheless, let's see what we can do.

*

This book sets out to explore just what it is that makes romantic relationships what they are. In the chapters that follow, I will explore the neurobiology and psychology of relationships, ask how romantic attachments differ from the other kinds of relationships we have (in particular between mothers and infants, friends and kin), and try to unpack something of their evolutionary origins. So let's begin with trying to understand the neurobiology that creates all the turmoil and intensity of falling in love.

2

Truly, Madly, Deeply

Ae fond kiss, and then we sever;
Ae fareweel, and then for ever!
Deep in heart-wrung tears I'll pledge thee,
Warring sighs and groans I'll wage thee.

'Ae Fond Kiss'

The poets speak of falling in love as though it was a kind of anguish – a sense of exhilaration tinged with loss, of yearning for what might have been. What on earth creates this extraordinary feeling? And why does it so often take us by surprise? One possible answer that got neuroscientists very excited a few years ago centred on the role of the neurohormone oxytocin. Its proper biological function originated in the management of water balance, but somehow during the course of mammalian evolution it became tangled up in the processes associated with reproduction, including both giving birth and lactation. It's not hard to see how a neurochemical implicated in managing water balance within the body should become involved in lactation. After all, that's just another form of water balance, since the water converted into milk has to be replaced to avoid the mother becoming too dehydrated. And then, perhaps, it's but a small step from that to the essential precondition for lactation, namely the birth process. It's yet another nice example of how evolution often exploits something that evolved for one function for some entirely different, but vaguely related, purpose.

Oxytocin really hit the headlines in the early 1990s as a

result of a series of studies on an obscure group of North American voles – tiny, mouse-like, burrow-living rodents that scutter about in the undergrowth. The key finding was that females of two species of vole that differed in their mating systems also differed markedly in the number of receptor sites for oxytocin in their brains. Although both species' brains released equal quantities of oxytocin, one species seemed to be much more responsive to it. Unusually for such a small mammal, that species – the prairie vole – also happens to be monogamous: after mating, the male stays with the female right through until the pups have been weaned some forty-five days later. The montane vole, whose females are less responsive to oxytocin, is promiscuous and the male does not stay with the female after they have mated. The rather too obvious conclusion was that oxytocin must be involved in the processes that underpin pairbonding behaviour, and in the popular press it was soon dubbed the 'monogamy hormone' – or the 'cuddle hormone' because it made it possible for voles (which are otherwise usually quite aggressive towards each other) to huddle together in their burrows. In short, oxytocin seemed to make a prairie vole female more tolerant of the continuing close proximity of the male with whom she had mated, to the point where she would allow him to share her burrow. Here, then, in one simple chemical process lay the elixir of life. Or so it seemed.

Love hormones?

The starting point for much of the interest in oxytocin had been the discovery that it was released in large quantities during mating, and particularly, in human females, during orgasm. Given oxytocin's role in the processes of birthing and lactation, it was, perhaps, not too surprising to find it also deeply involved in female orgasm. Its release at orgasm coincides with a great deal of mechanical stimulation of both the upper body (especially the breasts) and the reproductive tract, so it may be that it is all part and parcel of the same process of physical stimulation that triggers oxytocin release during birthing and suckling. Nonetheless, the fact that oxytocin is released in the wake of orgasm probably explains why many of the same sensations and emotions as are roused by birth and suckling also occur in this context. This apparent connection with sex led, naturally enough, to questions about its involvement in other aspects of romantic behaviour and pairbonding. If oxytocin is about bonding, then its job might simply be to strengthen the bond with whomever you happen to be involved with at the particular moment – baby in the first case, partner in the second. It's a cheap chemical trick to bypass your natural defences. Rational thought flies out of the window, and instead you get poleaxed whether you want to or not, your better judgement notwithstanding.

In rats, high doses of oxytocin give rise to a sedative-like effect, including lowered blood pressure and reduced locomotion. The physical stimulation of suckling itself (the process that triggers the release of oxytocin) is also associated with anti-stress effects, and there is evidence to suggest that oxytocin may be associated with physical touch: massage-like stroking of a rat's abdomen raises oxytocin levels and results in an analgesic effect associated with elevated pain thresholds. Similarly, in novel environments oxytocin-deficient female mice are more nervous and have higher physiological stress levels than genetically normal mice. These symptoms can be alleviated by injecting oxytocin directly into the rat's brain. Intriguingly, analogous effects have been reported for women: the frequency of hugs with the partner correlates with elevated oxytocin levels and lowered blood pressure (indicating greater calmness and reduced stress) during stressful situations. In addition, the release of oxytocin following natural birth is associated with changes in personality in women, including greater levels of calmness, sociability and – probably just as well – a greatly increased tolerance of monotony. It began to seem that in some mysterious way, oxytocin was involved in the very processes of social bonding itself. Whatever may be mediating this relationship (and it may actually involve other neurohormones, as I shall suggest below), it seems that hugs are good for you. In one study of couples, the best predictor of low levels of cortisol (the so-called stress hormone) was the frequency of intimacy: more time spent on intimacy had an immediate effect on cortisol levels each day.

Being willing to share a burrow or a home with a mate is, in the end, about trust, trust that they will not kill your pups (at least in the case of voles whose males have something of a bad reputation in this respect) or mistreat you. So it was perhaps no surprise when it was found that oxytocin also seems to play a role in facilitating trust even in humans: a whiff of oxytocin up the nose causes us to be more generous towards each other. This was demonstrated in a neat experiment in which subjects were asked to play the Trust Game. In this game, one player is given a sum of money and asked to share some, all or none of it with a second player. Whatever they give to the second player is doubled, and the second player is then invited to share some, all or none of the enlarged pot with the first player. The best strategy for the first player ought to be to hand over the whole pot to the second player. Providing the second player is honest and splits the enlarged pot with the first player, they both benefit to the maximum. But the risk is that the second player simply pockets the lot and does a runner. That way, player two gets a double bonus, and player one gets nothing. So most people try to hedge their bets and only give some of their pot to player two, keeping back a proportion for themselves so that at least they get *something* if it all goes pear-shaped.

But a whiff of oxytocin up the nose has a decidedly beneficial effect: on average, player one shared 17 per cent more of the initial pot with player two after a shot of oxytocin compared to when they were given a placebo spray with an inert chemical. What makes it clear that this is about trust is the fact that when the experiment was re-run with player one playing against a computer that responded randomly (and they *knew* they were playing against a computer), there was no difference between the oxytocin and placebo conditions in their willingness to share. In other words, it was not simply risk that player one was betting on, but malicious, sneaky, untrustworthy human behaviour.

Another study has claimed that oxytocin increases men's ability to correctly read social information in the eyes. However, the improvement in performance after a squirt of oxytocin up the nose was only a measly 3 per cent (something most people wouldn't consider biologically meaningful) and only worked at all in two-thirds of the men in the sample. Besides, the test of social skills used in the study wasn't, perhaps, the most sensitive: it is a test of knowing whether someone is looking directly at you rather than away from your eyes, and although a good test of autism (the psychological condition associated with the virtual absence of social skills), it isn't really a terribly good test of normal individuals' social or mindreading competences. Who knows what these results are actually telling us? More plausibly, however, in another study the same team found that oxytocin reduced amygdala activity in men when viewing photographs of faces exhibiting different emotions. The amygdala is normally involved in the management of our responses to negative emotions (fear, anxiety, anger), so it makes sense, given how it works in females, that it should dampen negative emotional responses even in men.

Although oxytocin is present in male brains, it doesn't seem to have quite the same dramatic effect that it has in female brains. Instead, the vole work suggests that males are responsive to a related but rather different neurohormone, vasopressin. Males of the monogamous prairie vole have more vasopressin receptors in the brain's ventral pallidum (part of the ancient limbic system whose main job is to manage our emotional responses), whereas those of the promiscuous montane vole have rather fewer. Remarkably, using viral vectors to manipulate the density of vasopressin receptors in the ventral pallidum directly affects the strength of partner preference in the absence of mating in male voles. Similarly, viral transplants of vasopressin genes into the ventral forebrain of male meadow voles (another promiscuous species) greatly increased the frequency of huddling with a female (i.e. pairbond-like behaviour). Intriguingly, vasopressin has also been found to have a significant role in social memory in male mice. They are more likely to remember whom they have huddled with previously when they have had a shot of vasopressin.

The fact that almost all the work on oxytocin and vasopressin has been done on rodents inevitably raises the thorny question of whether rodents are really an adequate model for humans. There have been major changes in sociability between the ancestral mammals (of which the voles are representative) and the lineages that gave rise to the monkeys and apes. Nonetheless, a recent study extended this line of enquiry to monkeys, bringing it slightly closer to home. Two brain areas specifically associated with oxytocin and vasopressin uptake (the nucleus accumbens and the ventral pallidum, respectively) were shown to be especially active in male titi monkeys when they were first paired with a novel female. There is, in addition, some evidence to implicate vasopressin in human behaviour. In a study of several hundred Swedish twins, men who had a particular (and relatively uncommon) version of the vasopressin receptor gene typically exhibited poor pairbonding. They were more likely to be rated as poorly bonded by their partners and more likely to have been divorced – despite the fact that everyone in the sample had been with his current partner for at least five years and had at least one child with her. The researchers concluded that they had found the gene for pairbonding in men.

So far so good on what seemed like a very neat story. A simple neuroendocrine mechanism was all that was needed to switch species from one form of mating system to another. Seemingly, it could turn aggressively promiscuous males into soft, cuddly pairbonders. Or a quick shot of vasopressin down at the local clinic, and you can turn your wayward cad into a home-loving, romantically engaged, everything-you-ever-wanted-from-a-lover kind of guy. Alas, the real biological world is seldom so simple. The spanner lurking in the works is the fact that, across the voles as a whole, the particular vasopressin receptor gene that was supposed to underpin the switch to monogamy does not correlate with monogamy at all. In fact, it turned out that the same gene occurs in almost every vole species whose genetics had been sampled, irrespective of whether or not it is monogamous or promiscuous (and all but a handful of the 155 species of vole are in fact promiscuous). In other words, it isn't the *presence* of an unusual vasopressin receptor gene in monogamous species that is unusual, but rather the fact that only a couple of promiscuous species don't have it! At this point, it was beginning to look like an unfortunate case of science by jumping without looking.

And there are other issues. In reality, this supposed mechanism for turning cads into dads seems to be rather short-lived in its effects. In the study of titi monkeys mentioned above, the vasopressin effect was only seen in newly paired males: males in long-term relationships behaved more like unpaired males. Similar results were found in a study of oxytocin in guinea pigs: the oxytocin effect wore off after just a couple of weeks. This may be fine for species like voles and guinea pigs that have very short-lived pairbonds. But it simply won't do for the much longer-term pairbonds we find in monkeys, apes and humans. It is almost as though the oxytocin/vasopressin system is an attention-grabbing mechanism: focus your attention on *this* one while you can. But it is an effect that wears off quite quickly. And a closer inspection of the Swedish twin study also raises doubts. It's as though the version of the vasopressin receptor gene possessed by the poorly pairbonded males simply influences the tendency to respond precipitously – to act first and think about the consequences afterwards, to take risks without thought for tomorrow. We'll see in Chapter 4 that risk-taking is a trait that males are particularly prone to, and that women may be especially attracted to risk-takers as casual mating partners. This apparent dichotomy between males who are well bonded and those who are not may actually have as much to do with different male mating strategies as with pairbonding as such.

And there's another odd thing about the oxytocin/vasopressin story when you stop to think about it. Why would evolution produce two completely different mechanisms for pairbonding in the two sexes? It doesn't really make any evolutionary or biological sense. Since both sexes produce both oxytocin and vasopressin, there really wouldn't have been any great problem about cranking up the same neuroendocrine mechanism for both sexes. If this really is all there is to pairbonded relationships, the fact that two separate neuroendocrines are involved would imply that the two sexes are radically different in their mating-game physiology, and so probably would not feel the same kinds of emotion in these contexts. Of course, that might well be true – after all, we can never really get inside someone else's head and feel what they feel, and men and women are always complaining that they can't understand each other (the *Men Are from Mars, Women Are from Venus* syndrome). But, even so, it isn't that obvious that the two sexes are *so* different in either behaviour or inner feelings, so we should perhaps have been a little more suspicious of the story.

The story gets more complicated

In fact, the business of pairbonding owes at least as much to a number of other neurotransmitters that are kicking around in the brain. One of these is dopamine. Dopamine normally plays an important role in facilitating the smooth execution of actions: people who lose the function of dopamine-producing cells in the brain suffer from Parkinsonism (poorly co-ordinated gait and an inability to control the movement of the limbs). It also seems to play an important role in reward. It seems that every time you see your lover, you get a shot of dopamine, providing very much the same pick-me-up effect as a shot of cocaine. Low dopamine levels, especially in the frontal lobes, have an adverse effect on attention and memory, and the presence of chronically low levels of dopamine in the frontal lobes tends to be associated with attention deficit disorder and the inability to concentrate and hold focus – in other words, just the kind of poor social engagement exhibited by the Swedish males with the unusual vasopressin receptor gene. So maybe it is dopamine and not oxytocin (or vasopressin) that gives you that electric moment when your eyes first meet across the room.

Another potentially important set of neuroendocrines that seem to be involved in social bonding are the endorphins. The endorphins are part of the pain control system; they are produced mainly by the hypothalamus, with receptors all over the brain. Endorphins are released in the brain in response to any stress on the body. Anything that induces pain or stress on the muscles triggers the release of endorphins, and that includes everything from jogging to serious circuit training and marathon running. Even psychological pain or the anticipation of pain releases endorphins. Endorphins build naturally to high levels in marathon runners in anticipation of a big meeting, as well as in women in the last trimester of pregnancy in anticipation of the impending birth. Indeed, it seems that it is endorphins that are responsible for the analgesic and relaxing effect of hugs, which had previously been claimed to be due to oxytocin: opiate antagonists like naltrexone block the effect in these cases, whereas oxytocin antagonists do not. In fact, it seems that the release of oxytocin triggers the release of endorphins in a kind of neurohormone cascade.

Monkeys and apes use social grooming to stimulate the release of endorphins. When monkeys groom each other, it is a bit like a massage. Folk wisdom tends to assume that grooming in animals has something to do with catching fleas, but free-living monkeys don't actually have that many fleas and other skin parasites – these tend to occur only when animals live in confined spaces (like our own houses) or under the protective cover of the nice warm clothes that we wrap around us. When monkeys groom, what they mostly remove is bits of vegetation that have become caught in their fur, or scabs and other blemishes on the skin itself. Grooming typically involves a gentle tugging and stretching of the skin. It's not unlike what you see mothers doing when they fiddle with their children's hair. In fact, it's really a form of massage, and the stresses imposed on the skin and muscles by massage are a very effective trigger of endorphin release in the brain.

We still do this kind of grooming activity, though we tend to call it petting and cuddling. More importantly, as with grooming in monkeys and apes, it is a very intimate affair, so we tend to restrict it to those with whom we have special relationships – our children, lovers, our closest friends. Too much physical stimulation in the form of stroking and massage with strangers, and it tends to flip over into an unintended sexual response. In part, it's probably a consequence of the intimacy created by the endorphin surge that comes from the mechanical stimulation of the skin. So we keep that mechanism for where it should belong – or, at least, where we want it to belong – namely, our most intimate relationships. But to get to this happy state of a 'very special relationship', we have first to bridge the gap we normally like to keep between us – our personal distance. Monkeys and apes who want to establish a new close relationship do this through a series of stages that involves presenting, touching, non-sexual mounting (a token mount, often by same-sex individuals, signalling a dominant–submissive relationship) and only later grooming. That endorphins play a role in romantic relationships has been nicely demonstrated in a study which showed that women in long-term relationships exhibited higher pain thresholds (indicating an active endorphin surge) when viewing a photo of their partner compared to viewing a photo of a male stranger or a pot, and also when holding their partner's hand behind a screen than when holding a stranger's hand or a ball behind the screen.

While the oxytocin surge may be common to all mammals from voles to humans, the role of endorphins in servicing close relationships over the longer haul seems to be unique to the primate family. This may be because the oxytocin surge (and perhaps even an associated dopamine surge) is quite short-lived, and its effects fade away within a week or two at the longest. This doesn't matter too much for voles, since their reproductive cycle is very short and the business of reproduction turns over very quickly. In any case, most voles only live for three or four months, at most a year. But, as we shall see in the next chapter, very early on in their evolutionary history monkeys and apes evolved a new, more strongly bonded form of sociality. This seems to have needed a more powerful bonding mechanism to support it, and the suggestion is that the endorphin mechanism was co-opted to provide the chemical underpinnings for this.

When it comes to bonding our relationships through the endorphin mechanism, we do have a bit of a problem. We live in super-large social groups where not everyone is as familiar with each other as they are in small monkey and ape groups. At one level, our solution to this problem has been to invent conversation. But conversation on its own is very dull stuff and hardly the basis for an intimate relationship. What we seem to have done is to use laughter to bridge the gap, because laughter turns out to be a very good releaser of endorphins. Laughter seems to produce a more generalised effect that applies rather more equally to everyone who happens to be in the conversation at the time, whereas physical contact is very much a one-on-one thing. Laughter allows us to trigger an endorphin effect in a less risky way.

Laughter, the best medicine

Laughter makes the social world go round. Conversations that lack laughter seem peculiarly dull, and just make us anxious to move on and find someone else to talk to. Although we share laughter with the great apes (it derives from their play-invitation behaviour), we have adapted it to occupy a central role in our interactions with each other. Tatiana Vlahovic and Sam Roberts, two members of my research group, recently completed a study of the role of laughter in interactions, and showed that we come away with a significantly higher level of happiness from an interaction if it involves at least some laughter than if it doesn't. They found that the occurrence of laughter (even in as symbolic a form as the acronym 'LOL'* or the smiley-face 'emoticon') has a much greater effect on how happy we feel after communicating with a friend than the amount of time actually spent interacting, irrespective of whether the interaction is face to face, by phone or via text messages or email. Amazingly, even just a simple 'LOL' in a text message can increase the happiness rating of an interaction by around 30 per cent.

The important element here seems to be relaxed or Duchenne laughter – named after the great nineteenth-century French neuroanatomist who identified many of the muscular processes underpinning facial expressions. Relaxed laughter contrasts with the 'polite titter' forms of laugh, which have more in common with nervous smiles and the bared-teeth expressions of submission that we find in monkeys and apes. Duchenne laughter leaves us feeling relaxed and genial, with the sense that all's well with the world. This kind of laughter is particularly effective at triggering the release of endorphins because all

---

* For those not in the know, LOL is an acronym standing for 'Laugh out loud', now commonly used in text messaging. An older generation more used to letter-writing probably assumes it means 'Lots of love'.

that heaving of the chest and abdomen is hard work for the muscles, and so endorphins are released to counteract the stress. It is not for nothing that we speak of laughing till it hurts. The light opiate high that we get from social laughter is not just rewarding – it makes us feel more trustful of, and more generous towards, those with whom we laugh. Indeed, in one of our experiments, it seemed to make us treat strangers as generously as we would treat friends.

I suspect that laughter is actually very ancient in its origin, not least because we share it with great apes. However, human laughter differs from great ape laughter in that it involves a much longer series of exhalations without intervening inhalations, and it is this that makes our laughter particularly exhausting. It is also more intensely social than is the case with apes, whose laughter episodes tend to be brief and fleeting. Human laughter is more like wordless chorusing, and my guess is that it started as a form of chorusing to reinforce grooming when there were more people involved in social interactions. Later, with the evolution of language, we found the perfect mechanism for triggering laughter and its accompanying endorphin release, namely the telling of jokes.

Testing for endorphins is a notoriously tricky problem, because endorphins (unlike oxytocin and dopamine) do not cross the blood–brain barrier, so you cannot measure them with a quick and painless blood sample. Your choices are either a lumbar puncture to obtain cerebrospinal fluid from the spinal column (neither the most pleasant nor the safest thing to do, since about 30 per cent of people who have a lumbar puncture have side effects of one kind or another[*]) or the form of brain scanning known as PET (positron emission tomography). PET involves injecting radioactive dyes into the bloodstream, and aside from the risks that it inevitably involves, it's very expensive, so this isn't a great deal better in reality. The simplest alternative way of assaying for endorphin release is to exploit the fact that endorphins are part of the brain's pain-control system, and to use a pain threshold test. If the pain threshold is higher after doing an activity than it was before, then that implies endorphins have been released.

We have used changes in pain threshold to explore the relationship between laughter and endorphins in a number of contexts. Most of these involved subjects watching videos of either stand-up comedy or boring documentaries ('how to play better golf' programmes are my favourite for the latter category) and testing their pain thresholds beforehand and afterwards. After laughing at stand-up comedy, pain thresholds are invariably higher, whereas there is no change in pain threshold after watching boring documentaries (indeed, if anything, they are actually lower afterwards). This suggests that it is the act of laughing that is crucial here. One of my students, Rebecca Baron, took the test to the Edinburgh Fringe Festival and ran it in real life at a set of stand-up comedy events and some short drama performances. Once again, the comedy produced the usual

---

[*] You are, after all, jabbing a relatively massive needle (it has to be large in order that it doesn't snap off under the force you have to use to get it through the cartilage between the vertebrae) into a direct pathway to the brain. Basically, it's a perfect way of introducing infection straight to the brain as fast as possible – possibly not the most sensible thing to do without good reason, even in the interests of science.

increase in pain threshold, but the drama (at which there was no laughter) did not.

So laughter seems to be important in regulating our social relationships, and it does so through its ability to trigger the release of endorphins. This makes us feel warm and positive towards those with whom we do this curious activity. And that is no doubt how it then came to play an important role in the processes of courtship. Laughter, through the endorphin high, allows you to trust a stranger, and that opens the way to get to know them better. Gradually, step by step, encouraged by the continuing stream of jokes and one-liners, the witty remarks and humorous asides, you are drawn into the spider's web.

The whiff of success

Getting up close and personal opens up new ways in which your rational judgement can be undermined by yet more cheap chemical tricks. And smell is one of those. It's not for nothing that the perfume industry is worth billions of dollars each year. Our perfume preferences are very personal: the ones we like to drench ourselves in are those that seem to be directly related to our own natural body odour. In effect, we prefer to wear perfumes that enhance our natural smell rather than ones that cover it up, which is why it's always tricky buying perfumes for someone you don't know extremely well. Nonetheless, it turns out that, if you can get in close enough, checking out someone's smell is a valuable way of finding Mr or Ms Right.

But before we get into this, let me briefly digress into a cautionary tale. If there is one thing the Eskimos are famous for, it is surely the fact that, when they meet, they rub noses instead of shaking hands like we do. Right? Alas, this turns out to be just another of those old travellers' tales that is at best a half-truth. When European explorers first came across the Eskimos, they observed what seemed to them this rather strange ritual. Trying as best they could to make sense of it, they interpreted it as a rubbing of noses. In fact, Eskimos don't rub noses at all. What they do is place their noses close up to each other's face and breathe in deeply. We find exactly the same behaviour among the Maori peoples of New Zealand, who refer to this behaviour as *hongi*. In their case, it involves a light press of one nose upon another in a symbolic joining together of host and visitor. But the origin of it, and in many ways its whole essence, is about breathing in the spirit of the other person.

In both cultures, what they are actually doing is breathing in each other's smell and, in effect, identifying each other. In fact, it is not all that different from the southern European 'air kiss' that has become ubiquitous in the past decade or so. We northern Europeans are so used to pecks on the cheek that we make the mistake of still trying to give a kiss, when in fact we should just be breathing in deeply. Another context in which this is much more obvious is when people greet babies. Watch next time someone holds a baby or toddler for the first time, especially if they are related. They will lift it up so the baby's face or chest is close to theirs, close enough to inhale its scent without it being too obvious. Actually, some people are very open about it, taking a deep breath when they do this. I have even heard people comment on how much they love the smell of new babies. No, you don't! You're just making excuses so that you can check out its smell.

A curious story attached to the great Mongol leader Genghis Khan provides another relevant example of how important smell is cross-culturally. The thirteenth-century *Secret History of the Mongols* records that Genghis's father, Yesugei, was out hawking one day when he came across a cart bearing a young woman accompanied by her husband. Yesugei recognised them from their clothing as Ongirads, another of the Mongol clans. After catching a glimpse of the young woman, Yesugei was overwhelmed by her beauty. Having gone to fetch help, he proceeded to chase and harass the party in an attempt to capture her. The woman, whose name was Chiledu, begged her husband to flee. As he did so, she ripped off her shirt and threw it to him, crying, 'While you live, remember my fragrance!' He escaped, and she never saw him again but instead became Yesugei's wife and soon afterwards the mother of Temujin, the future Genghis Khan.

Despite our aversion to smell and our much reduced olfactory areas in the brain (at least compared to dogs and horses), we are in fact surprisingly sensitive to it. Newborn babies and their mothers can identify each other by smell alone within hours of the birth – which is one reason why we now like to make sure that the baby goes straight on the mother's breast as soon as it is born. This is something that we share with most other mammals. In sheep and goats, the mother learns to recognise its newborn young by smell within twenty-four hours, and in the following days will only allow that lamb to suckle. And the lambs themselves learn to identify the right mother to suck from in the same way, though they are, perhaps understandably, a bit slower and it usually takes a couple of days' exposure to the mother's smell.

In fact, smell provides one of the best markers of who you really are. The reason for this is that your smell is determined by the same set of genes, the major histocompatibility complex genes (MHC), as your immune system. It is part of who you are, your personal chemical signature. The MHC gene complex is particularly susceptible to mutation, producing new immune complexes with each new generation. This is probably just as well, as these are our first line of defence against bacteria and viruses which are themselves undergoing constant genetic change. Our immune-system genes have evolved to be almost as changeable as virus genes in an effort to track the ever-changing biological threats that we face from them.

So smell may be one way of checking out who's a good bet and who's not, but it's not the only function of smell in this context. Female moths famously dribble molecules of an incredibly powerful scent into the air. Male moths can detect these scents in the tiniest quantities from hundreds of yards away and find them quite irresistible. These sexual attraction scents are known as pheromones and occur widely in the animal kingdom, including in monkeys. There has been some debate as to whether or not they occur in humans, but in fact there is considerable evidence to suggest that they do.

There do appear to be significant differences between the sexes in their respective sensitivity to odour: women are much more sensitive than men. There is now quite a lot of evidence that women, in particular, are quite good at identifying their children and their lovers by scent alone. However, we are by no means perfect at this, it must be said, and it is probably just as well that we don't manage our social world by smell rather than by vision – we would be likely to make an inordinate number of embarrassing mistakes if we did. However, it seems that, having identified the right person, smell plays a very important role in sexual arousal for women in a way it doesn't for men. Perhaps as a result, women rate smell as more important in mate choice than men do, whereas men rely much more on visual cues, reflecting the fact that men tend to make up their minds about a prospective mate from further away than women do. Women need to get up close and personal. In a large questionnaire-based study, Jan Havlíček, Tamsin Saxton, Craig Roberts and their colleagues found that women rated odour as more important than visual cues in a range of non-sexual contexts (such as meal choice, flower choice and attention to unfamiliar landscapes) as well as in contexts of sexual arousal and lover choice, but men did not. For men, visual cues were much more important, especially in sexual contexts and lover choice.

Some years ago, Kate Willis, then one of my students, ran an experiment to determine whether men could tell when a woman was ovulating by smell alone. Six women each wore a T-shirt on three successive nights during each of the four weeks of their menstrual cycles. At the end of each week, eighty men were them asked to sniff the six T-shirts and rate them on a simple scale of pleasant-to-unpleasant. To avoid contamination, the women had to be non-smokers, and they had to avoid highly spiced foods and using scented soaps or perfumes or hormonal contraceptives while they were involved in the study. The results were very clear: T-shirts that had been worn around the time of ovulation were rated by the men as being significantly more pleasant than those worn at other stages of the menstrual cycle, and as more pleasant than those worn by post-menopausal women or women who were on hormonal contraceptive pills. It seems that, in some indefinable way, men are more attracted to women when they are ovulating, and can in effect detect when ovulation is occurring. Or, to put it the other way around, women use olfactory signals to entice men into coming closer when they are ovulating.

Such effects work both ways, of course. Androstenol is one of a family of steroids formed as a natural by-product of testosterone, the so-called male hormone. It's responsible for the slightly musky smell that men naturally have, and is one of the components of truffles. In an infamous experiment, three psychologists, Gustavson, Dawson and Bonett, once sprayed androstenol around half the cubicles in men's and women's toilets. Then the researchers recorded how often users who had a free choice of all the cubicles (i.e. none were occupied) entered the ones treated with androstenol. What they found was that men tended to avoid the androstenolised cubicles – having ventured in, they would usually back hastily out and find an androstenol-free one instead. But women apparently found the androstenolised cubicles rather congenial – even if not irresistible – and used them more often than the untreated ones. In contrast, when the same cubicles were later sprayed with a related by-product of testosterone produced in the liver which serves very different physiological functions from androstenol, neither sex exhibited any preference.

In an updated version of this experiment, Tamsin Saxton and her colleagues at Liverpool University applied androstadienone (another of the same family of testosterone-derived steroids) to the upper lips of women at a speed-dating event. In speed-dating (for those of you who have yet to experience this novel form of mating market for the ultra-busy), the women sit round the room at tables and the men spend five minutes with each one in turn, moving on one place when a bell is rung. At the end of the evening, everyone lists the names of the people he or she would like to meet again, and the organisers then exchange details for them. In this study, the androstadienone was concealed in clove oil to disguise it. To control for the effects of other odours, a third of the women had androstadienone plus clove oil, another third had just clove oil and the final third had plain water. That allowed the experimenters to separate out the effect of the clove oil substrate from the androstadienone itself. The results could hardly have been more conclusive. The women who had received the androstadienone not only rated the men they met at the speed-dating event as more attractive than did the women in the other two groups, but they were also significantly more likely to ask to see them again. Somehow, the androstadienone acts on deeply buried brain mechanisms to create a rosier-than-reality view of the hulking brute before you.

In a variation on this, Angeliki Theodoridou, Ian Penton-Voak and colleagues gave subjects either a single dose of oxytocin through a nasal spray or a placebo control that had all the same ingredients of the spray except for the oxytocin. Then they asked them to rate the trustworthiness and attractiveness of photographs of nearly eighty different faces, half of them male and half female, all posed with a neutral expression. Those who had had the oxytocin spray rated the faces as more trustworthy *and* more attractive than those who had had the placebo control. Who said romance was dead?

And this perhaps explains why a good sniff of a strong perfume or aftershave close up can sometimes turn your head.

Ae fond kiss?

But smell is only as good as a washed body. Smells can be masked, not just by ladling on Givenchy's latest, but also, in the state of nature in which we have spent most of our evolutionary history, by accumulations of dirt and bacteria. One way to circumvent this problem is to get up even more close and personal and taste the stuff directly. In mammals, saliva is full of chemicals exuded by the body, not the least of which are a group of proteins known as major urinary proteins, or MUPs for short. The name comes from the fact that they were first identified in rodent urine, where they play an important role both in allowing individuals to recognise each other and in regulating territorial behaviour. Female mice can discriminate between males solely on the basis of differences in their MUPs. That way, they know whose territory they are about to enter, and can choose between males to get the best of the bunch. Not all mice MUPs are used in individual identification, however. One particular MUP seems to have the same form in all males, and is unique to male urine – it is never found in females. And females find it *very* attractive, so it seems to be a kind of mouse pheromone, a scent designed to attract members of the opposite sex (rather like androstenol in humans). Because female mice find it so irresistible, this particular MUP was given the name *darcin*, in explicit honour of Mr Darcy in Jane Austen's novel *Pride and Prejudice.*

Freud, who was famously hung up on childhood complexes, took the view that kissing has something to do with the memory of the pleasures of sucking on your mother's breast. I guess it's easy enough to see how he made the connection, but the argument really doesn't hold water. After all, if it really is a reversion to breast-sucking, why not just do *that*? Another suggestion, this time from ethologists, is that it's a form of courtship feeding, a habit widespread in the insects and among some birds. But even then, the argument doesn't really add up. Courtship feeding is a decidedly male thing, with males offering packets of food (sometimes regurgitated, sometimes not) as gifts to prospective mates. In the species that indulge in this kind of thing, the food package provides the female with essential nutrients that she needs to produce a decent batch of eggs. Females judge the quality of a male by the size of his offering. But kissing doesn't actually provide anything besides a little spittle, and that can hardly be considered crucial to a woman's ability to produce eggs. Besides, both sexes do kissing with equal enthusiasm, and true courtship feeding is usually a one-way trade (or at least a trade in which the reciprocal benefit comes later and in a very different form, namely sex). In any case, if it's a matter of providing resources, we do all this anyway in the rather more useful form of offering lavish gifts, not to mention the salary cheques that help pay the rent and put dinner on the table. And if you want food offerings, well, think box of chocolates, perhaps even flowers, and certainly invitations to dinner. Courtship feeding cannot be what kissing is about. Something else must be afoot.

In fact, forget about courtship feeding, and forget about Freud: the answer lies in a completely different direction. Kissing is probably all about testing the health and genetic make-up of prospective mates. Health is perhaps an obvious one, because poor health is reflected in bad breath and a sour taste in the mouth, and these are very easily detected when kissing. In the previous section, I discussed the importance of our immune system and the MHC genes that underpin it, and how these are related to your own personal smell. Research from a number of labs suggests that we choose our mates as a function of how their MHC complex compares with ours. Because the MHC genes determine your immune responses, we tend to prefer people who have a different set of MHC genes, presumably because doing so allows us to produce offspring with a wider complex of immune responses. So a good kiss, as much as a good sniff, not only allows you to check out how closely someone is related to you (related people will have smells in common) but also who isn't related to you and so might make a good mate. In a study of forty-eight romantically involved couples, the women rated their sexual interest in their partner as significantly greater when they shared fewer MHC alleles than when they were more similar. As the proportion of shared alleles increased, their sexual responsiveness to their partner decreased, the number of extra-pair sexual partners they had increased, and so did their attraction to other men.

In a way, it was just an accident of history that MUPs were first discovered in mouse urine. In fact, they occur in urine simply because urine is a very convenient mechanism for rodents to deposit signals of their presence round their territory, and the researchers who discovered them just happened to be interested in how the animals marked and defended their territories. In many other mammals, MUPs are found in saliva. Although humans lack the genes for MUPs as such, their saliva contains a very large number of proteins of a broadly similar kind – 1,116 at the last count. Most of these are involved either in protecting us from toxins and bacteria, carcinogens and other poisons, or in aiding digestion by helping to break down the food we have eaten. About a quarter of them are also found in our tears. Individuals seem to differ a great deal in the particular collection of proteins they produce, and although so far there has been no interest in exploring the role of saliva in mate choice, it's there and is as perfectly designed for the job as MUPs.

Whether or not MUPs or the MHC genes are involved, kissing – and I don't mean a coy peck on the lips – provides a great deal of other information about a prospective partner, and there are striking sex differences in how kisses are evaluated. The results of a large survey study carried out at the State University of New York at Albany revealed that, as one might expect given that they are chemically more sensitive than men, women are more likely to use taste and the smell of the breath during kissing in evaluating a partner's kissing ability, and to use this to evaluate the quality of a male. They are also less likely than men to have sex without kissing first, and less likely than men to have sex with a bad kisser. Women thus seem to place more reliance on kissing for evaluating prospective partners.

Kissing, of course, isn't just about mate choice – otherwise we wouldn't bother to do it any more after we have tied the knot. The same study showed that kissing also plays an important role in arousal and eliciting willingness to have sex, especially in women (one likely reason why prostitutes typically refuse to kiss their clients). Women are also much less willing to engage in pre-coital kissing in short-term relationships or when the partner was only interested in sex than they are in long-term relationships. It is possible that saliva contains hormones that trigger sexual arousal, and, if so, one of the secondary functions of kissing might be to facilitate the passing on of these hormones from one sex to the other in order to bring both sexes into the same pitch of readiness to mate. Indeed, it is even possible that kissing has slightly different functions in the two sexes: social bonding for women, but facilitating sexual arousal in the partner for men.

In real life, we are, of course, only dimly conscious of these processes at best. It's probably just as well, because when we try to behave consciously in these kinds of context, we invariably mess it all up. Much of the processing of the cues we gain from kissing goes on well below the horizon of consciousness. So there probably isn't much to be gained from a detailed clinical understanding of the processes involved. Seems like the best thing to do is just lie back, switch off and let nature take its course . . .

*

In this chapter, I have focused mainly on the neurochemical mechanisms that underpin romantic relationships. How the brain processes all these complex signals, however, is still something of a mystery. The one thing we do know is that our massive brains evolved specifically to deal with the complexities of our relationships. In the next chapter, I take up this side of the story and ask whether our brains are designed to make us monogamous.

3

The Monogamous Brain

> John Anderson, my jo, John,
> When we were first acquent,
> Your locks were like the raven,
> Your bonnie brow was brent;
> But now your brow is beld, John,
> Your locks are like the snaw;
> But blessings on your frosty pow,
> John Anderson, my jo.
>
> 'John Anderson, My Jo'

Although we know exactly what a relationship is when we see it or experience it for ourselves, we find it genuinely difficult to say exactly what we mean by it – one reason perhaps why we admire poets so much for being able to capture in words something that we cannot ourselves express. Our difficulty seems to derive from the fact that a relationship is something we feel, rather than something we actually think about. Yet falling in love is the most important decision we can ever take – a passport to lifelong bliss (well, in theory anyway). At the outset, when everything seems rosy and perfect, we think the relationship will last forever. For some it does – they remain together until, proverbially, death does them part. But at the other end of the spectrum, for some people life is one long succession of failed relationships. Why do some people seem to be so much better at choosing a mate than others?

Even though we do sometimes devote a great deal of time and effort to thinking about whether to offer or accept a marriage proposal, few of us have probably gone quite as far as Charles Darwin in making a list of reasons for and against marrying his cousin, Emma Wedgwood. Nonetheless, we probably do spend quite a lot of time asking ourselves about our relationships. *Why did he say that? What did she mean? What if . . . ?* One benefit of having a big brain is that you can think about the implications of your actions, assess how another person is likely to react, ponder the reasons why someone behaved the way they did, and even occasionally ask about your own motivations (although, on the whole, as Freud famously noticed, we tend to prefer not to do *too* much of that).

We can do this only because we have brains that are, relative to our body size, much larger than those of any other mammal species. Indeed, our brains are actually absolutely bigger than those of all other species with the exception of only a handful such as whales and elephants. Even the giant dinosaurs had smaller brains than we do. Mind you, sheer size isn't everything. Although whale brains are larger than ours, their brains are organised rather differently: they have one less layer of neurons in the cortex – that crucial outer stratum of the brain where all the smart stuff goes on – and that has significant implications for their cognitive abilities. Nonetheless, the bottom line is that both our brains and theirs are designed to manage our relationships. Our brains provide us with a semi-rational computer that complements the purely emotional component arising from the neuroendocrine systems that we discussed in the last chapter.

Just how demanding can pairbonds be?

Although they are small, neurons are very costly to produce and maintain. Relative to their size, they cost about ten times more energy just to keep ticking over than the average for the body as a whole, never mind what happens when they are actively firing away as we think deep thoughts. Keeping a nerve in readiness so that it can fire when nudged by a signal from another nerve is massively expensive. In addition, there is the cost of replacing the neurotransmitters every time the nerve has fired. Monkeys and apes pay this price in order to be able to live in large groups. In this family of species, the average size of social group within a species is directly related to its brain size, in particular the size of its neocortex (essentially the thinking part of the brain): species that live in big groups have big neocortices. When Susanne Shultz and I began to look beyond the monkeys and apes (generally considered the smartest of all the animal families) to see how other mammal and bird families compared, we had expected to find the same relationship as in the primates: that brain size increases in step with the typical size of species' social groups. To our surprise – and consternation, initially – it did not. There simply wasn't a relationship between brain size and group size in any of the mammal or bird families that we looked at. Instead, it turned out that the species which had the largest brains all lived in monogamous pairbonded social systems. This was true for each of the three largest families of mammals that we examined (the carnivores, the ungulates and the bats) as well as for the birds as a whole.

At first, we were puzzled by the fact that it was the

monogamous species that had the biggest brains. Everyone had assumed that tracking the many relationships each individual has in polygamous or promiscuous mating systems would be much more demanding. After all, not only are the males mating with many females, they are also having to battle it out with all the rival males in the community, and it seemed plausible that knowing who these males are and how you stand with respect to each of them in the dominance hierarchy ought to be useful. And on the human front, one might imagine that trying to handle several co-wives at the same time, or, worse still, trying to prevent two lovers finding out about each other, must surely be extremely challenging. Yet our data clearly suggested that species with the supposedly simplest of all mating systems – monogamy – seemed to pay the highest price in cognitive terms.

Eventually, we were forced to consider the possibility that a monogamous pairbond might actually be much more psychologically demanding than any number of casual relationships. The essence of the problem is summed up rather neatly by the birds, most of whom in fact opt for some form of monogamy. About 85 per cent of bird species adopt pairbonded monogamy because the business of feeding nestlings is energetically very demanding, as well as being time-consuming. The breeding pair have to work their socks off to bring enough food back to the rapidly growing nestlings. They are willing to do this because it allows them to produce large-brained offspring. Large brains are very beneficial for birds: they allow them to be ecologically more flexible, to forage in more innovative ways, and to cope with greater seasonal changes in temperature and foraging conditions. For species that breed at high latitudes, it means they don't have to migrate to the tropics every winter to escape the challenging conditions that winter brings. Migrating, as most of the smaller-brained species do, certainly gets you out of a winter environment problem, but the business of migrating is itself fraught with life-threatening risks that a big brain doesn't necessarily help you with – the risks of being blown off course, of starving on the way, and of being caught by predators in environments where you are unfamiliar with the location of refuges or feeding places. Big brains allow bird species to stay in northern latitudes throughout the winter and so to avoid these risks. In turn, this makes them more resilient and demonstrably less liable to become extinct.

The cost they bear for doing this, however, is the effort involved in rearing chicks with big expensive brains. One parent can of course do the job of rearing young successfully, but what it sacrifices by doing this on its own is both the number of offspring it can rear and the size of brain its offspring can have. And that in turn inevitably means reduced ecological flexibility and having to find other ways to survive (like migrating when times get hard) – so it's a vicious circle and a bit of a non-starter. If the breeding pair work together, they can produce bigger-brained offspring and avoid these problems.

The emphasis here is on the working together bit. The pair have to maintain a conveyor belt of grubs and insects for the nestlings. If one of them slacks or dawdles in the spring sunshine, the other must bear a disproportionate share of the burden if they are to finish the job and successfully rear the chicks. The chicks can't wait! This is even more obvious during the incubation period, when the eggs in the nest have to be kept warm by someone sitting on them. The pair have to take it in turns to incubate and to feed. There isn't the option for one member of the pair to wander off to feed and then stay away all weekend down at the avian equivalent of the pub. A small finch-sized bird has to eat almost its own body weight in food each day just to stay alive. If its mate does not come back and relieve it in reasonable time, it will be left with the invidious choice between abandoning the eggs in order to feed or staying on the nest and starving. Either way, the likelihood is that the offspring won't survive: if cooling doesn't kill them in the first case, nest predators certainly will in the second.

So it looks as though it's the co-ordination and synchronising of activities that is so psychologically demanding for pairbonded species. In effect, each member of the pair has to be able to factor its mate's needs and requirements into how it plans its own day. Even a humble bird has to be sensitive to the day-to-day changes in its mate's behaviour in a way that promiscuously mating species simply don't need to be. I first came to appreciate this when I was studying the miniature klipspringer antelope in East Africa. In this species, the male follows his female around whenever she moves. He is seldom more than a few feet behind her, and never more than ten yards away. If you find one, you're sure to find the other nearby, probably watching you from under a bush. Because they are small and live on exposed rocky cliff faces, they are especially exposed to the risks of predation. To reduce those risks, they almost never feed at the same time. One will keep watch while the other feeds. When the feeding animal has had its fill and wants to rest so as to ruminate and chew the cud (an essential process in the digestion of grasses and other vegetation for members of the deer–cow–antelope family), it will come and stand by the side of the mate on guard, and may even gently nudge it on the shoulder as though to say: OK, your turn now. The mate will then move off and start to feed, while the previously feeding animal will settle down to keep watch and chew the cud.

If eventually they both settle to rest, it doesn't matter too much, of course, since both are able to remain alert to the possible approach of predators and other threats. Usually, they do this by lying facing in different directions so that they can cover all eventualities should a predator try to approach them. Because the female has the higher energy burden thanks to the costs of gestation and lactation, she will usually be the one that wants to start feeding again first. When she does get up and go, the male rarely waits long, but is up and after her within at most half a minute, tracking her every move until she eventually finds a spot where she wants to start feeding, perhaps a hundred yards away from where they were resting. As soon as she has settled to feed, the male is back on guard again and will stand alert, often on a small rock that gives him a good view, until the female wants to ruminate. Then he will have his turn to feed.

These endearing little antelope, no bigger than a lamb in size, must be the most obligately pairbonded of all mammals. They simply do not leave each other's side. There is considerable evolutionary pressure on them to behave in this way, because if one of them loses its mate to a predator, the chances of finding another are limited. In this kind of obligately monogamous world, everyone is spoken for, and there is no surplus of unpaired individuals waiting in the wings from whom one can pick a new mate. In any case, even if they do find another mate quickly, it is likely to be a young animal with little experience of the world. These young animals are still naïve and less clued in to the need to share duties. They will tend to wander off and start feeding again instead of waiting for the mate to finish – often forcing the more experienced mate to sacrifice feeding time in order to ensure that someone is on guard. So there is considerable pressure for them to look after and nurture the mate they've got because that's pretty much the best they'll get.

This raises another reason why monogamy might be psychologically taxing. Conventional wisdom among evolutionary biologists has been that if you are stuck with one mate for life (or, at least, more or less for life), then you had best not be too cavalier in your choice of mate by taking the first one that hove into view. It might be wise to size up the options and try to choose the best one you can. Presumably, the best in this context means the one that will live at least as long as you will, has a decent set of genes on offer to mix with yours for that perfect offspring, *and* won't wander off and spend too long away down at the pub leaving you to starve on the nest . . . or, in the case of the klipspringer and other similar species, won't wander off and start feeding while you've got your head buried deep in the grass and can't see the predator creeping up on you.

Choosing the perfect mate is no mean task when you have to try and balance many different dimensions, all of which might be important. The real problem is that the penalties of getting it wrong are massive. In the case of lifelong monogamy, making a bum choice can, to put it bluntly, be a disaster, and never mind just on the personal level. Choose an infertile mate, and your fitness – your contribution to the species' future gene pool – instantaneously drops to zero. Or, again, choose one that is less than assiduous in sharing the burden of successful reproduction, and you will do much less well in the evolutionary race than you might have done had you chosen more wisely. The costs of a poor choice are so enormous that they must have placed – and continue to place – enormous evolutionary pressure on animals to fine-tune their choices, to be sensitive to the cues that indicate a dodgy wastrel or a simple incompetent.

Plausible as this argument seems, there is a fatal flaw in it, and this is that it applies equally to species which mate promiscuously. You don't need to be stuck with the same mate for life to have to worry about choosing wisely. Even those who find a new mate each breeding season – the annual pairbonders, like many of our small garden birds – need to worry about avoiding the duffers. Of course, it's not quite such a disaster if you can find another mate next year, but a long succession of duffers won't do your biological fitness any good. Even among the promiscuous maters, choosing a series of poor-quality mates who have less than ideal genes to offer isn't the sharpest thing to do. So, as plausible as this argument seems, there is an element of doubt as to how it can really explain why lifelong monogamists need to be so much more careful than anyone else.

But there is another reason why this argument doesn't hold as much water as might seem reasonable at first sight. For many decades, we have assumed that the brain, once past the early stages of development, is more or less fixed.

If you lose a bit of it later in life, that's it for whatever function it underpinned. The famous case of poor Phineas Gage is always trotted out at this point. The handsome and rather debonair Gage had been the foreman of a construction gang building the new railway near Cavendish, Vermont, in the north-eastern USA in 1848. One day, while preparing an explosive charge to blast away some rock that was obstructing a cutting, he accidentally ignited it with the metal rod used to pack the explosives into the holes drilled into the rock. The tamping iron went up under his left cheekbone, through his left eye, and took out most of the left frontal lobe of his brain.* Though he survived and lived for many years, Gage famously lost all his social skills. Having previously been a particularly reliable foreman whose people skills were widely respected, he rapidly became an unpredictable, erratic, cussing-and-swearing drunk, prone to gambling. His career fell apart, and in the end he died in poverty twelve years after the accident.

To be fair, poor Phineas did lose a rather large chunk of his brain, and in a bit of a critical region for social skills, so it may be no surprise that his brain couldn't compensate by transferring the functions involved elsewhere. Nonetheless, it has turned out that even adult brains are more flexible than we had originally imagined. They are not infinitely flexible, of course, but there is evidence of some capacity for units in the brain to grow *and* regress (get smaller) as a result of circumstances. One of the

---

* You can see a reconstruction of this in an article by Peter Radiu and Ion-Florin Talos in the *New England Journal of Medicine* (2004, vol. 351, article e21), including a graphic video reconstruction of the tamping iron passing up through his skull.

earliest pieces of evidence for this came from studies of birds like the North American chickadee, which habitually caches seeds during the autumn for retrieval during the winter when food is scarce. Such species exhibit a seasonal growth and regression of the hippocampus, a tiny part of the brain involved in creating the spatial maps that we use for navigation – and, of course, finding things we've previously hidden.

Later, an inspired study of London taxi drivers by Eleanor Maguire showed that they had an unusually large hippocampus (or, at least, one particular lobe of the hippocampus) compared to normal individuals. London black-cab drivers are unique because they have to undergo a four-year training in which they essentially learn the map of the city's rather convoluted street system (or, should we say, lack of system). They can only become taxi drivers once they achieve a high enough proficiency on this to pass the regulatory tests. Subsequent work by Maguire's group has suggested that working taxi drivers have a larger than average hippocampus because it has grown with use, not because the task is self-selecting (as would be the case if, of all those who go in for the training, only the ones that have an unusually large hippocampus survive to pass).

All this simply goes to show that the brain can adapt and maybe even regenerate itself in the face of life's exigencies. So if birds and taxi drivers can adjust the size of bits of their brains, why on earth can't species grow and regress those bits of the brain needed for sensible mate choice according to whether they need them or not? Lifelong pairbonders should be able to dispose of all this expensive extra brain once they've found a mate, and settle down to a nice, quiet, uncomplicated life on half the

called The Knowledge

brain power. So they ought to end up with *smaller* brains than annual pairbonders (who, after all, have to make a choice every year), or, worse still, promiscuous species who make these decisions many times in the same year. Yet brain size goes in exactly the opposite direction. There must be something more to it.

Our research suggests that this something is the continued demands of maintaining the pairbond through time, including co-ordinating and synchronising one's behaviour with that of one's mate. In effect, there is something about maintaining a pairbond over the long term that is especially demanding. In the end, your partner has the whip hand: if they don't like your behaviour, they can always leave and find someone else. So your problem is to maintain the mate's commitment to you.

Scheherazade triumphant

In his book *The Mating Mind*, evolutionary psychologist Geoff Miller suggested that the need to be continually advertising one's value as a mate might explain the evolution of language and, through this, the evolution of the unusually large brains we humans have. He termed this the Scheherazade Effect, in reference to the heroine's linguistic skill in keeping King Shahryar* entertained in the ancient Persian story *One Thousand and One Nights* (sometimes also known as the *Arabian Nights*). By telling a never-ending story, she managed to avoid the fate that had befallen all her predecessors, namely execution the following morning for boring the great man. Miller thought that sex differences in the strength of this effect might provide an

* Actually, *Shahryar* just means 'king' in Persian.

explanation for other aspects of our behaviour in terms of sexual selection. One such compelling argument was the marked sex difference in language, such as the much larger, more flowery vocabularies that are especially characteristic of males, and their more boastful styles of speech compared to women's. This reminded him of the peacock's long train and elaborate mating displays. Human males, he thought, might display their qualities to passing females with their verbal skills and tales rather than the more conventional tails.

It seems to me that there is some merit in this suggestion, though I think the story that Miller originally presented to account for the enormous size of human brains probably confounds two equally important, but very different, processes. One is mate advertising, for which a flowery vocabulary and verbal eloquence may well be a good cue of genetic quality, if only because it shows what a good brain you have. The other is maintaining the relationship once the knot has been tied – and that might well have a lot to do with some kind of entertainment value, a genuine Scheherazade Effect. The question is: which of these two is the neuronally more expensive?

But there is another crucial issue hidden beneath all the male's flowery rhetoric. Strictly speaking, a sexual selection argument implies that the effect is acting differentially on the two sexes. Sexual selection is a very powerful force in nature and has been responsible for many spectacular examples of sex-biased traits, most of which are peculiar to males. Among the better known are the peacock's tail and the bird of paradise's elaborate courtship rituals, as well as the red deer's horns and human beards. This is because sexual selection (at least in this form) normally involves one sex (the 'coy' sex) having the main say in who gets to mate with them, and the other sex having not much to do except advertise its wares in the hope that some passing female might condescend to mate. Because of the way reproductive biology works (especially in mammals), it is the females that ultimately have most at stake from botched choices: they are the ones who are left to carry the can, and so it is they that ultimately have the whip hand and are able to exert pressure on the males. However, sexual selection rarely produces traits *de novo* out of nowhere. In most cases, it takes an already existent trait and wildly exaggerates it. So if language now functions in this way as a mating display, it is almost certainly because it already existed in some more mundane form to do another kind of job. In my view, that job is, and has always been, servicing social relationships. On this broader interpretation, the Scheherazade Effect has, I think, much to recommend it as a kind of iconic descriptor of what's involved in the long process of relationship maintenance.

Although it is perhaps overly simple to suggest that keeping a relationship going is just a matter of ensuring one's mate is entertained, there is a sense in which entertainment really does play an important role in oiling the wheels of relationships. It's the all-important role of laughter that I discussed in the previous chapter. Just how important a sense of humour is to us in our relationships is evident from the frequency with which 'GSOH' ('Good Sense of Humour') is used both as a requirement and as an advertised trait in lonely hearts columns. In just over a thousand heterosexual personal adverts that we sampled from the *Observer* newspaper in the late 1990s, a sense of humour was one of the most commonly listed of all types of traits. Of women advertisers, 48 per cent mentioned that they had a sense of humour and 53 per cent wanted a prospective partner to have one, while the equivalent figures for male advertisers were 45 per cent and 41 per cent. Women evidently placed a higher premium on a sense of humour than men did, but both seem to view it as an important requirement in a relationship – often more important than all other traits.

I have no doubt that the Scheherazade Effect plays a significant role in servicing our romantic relationships, but it almost certainly isn't of itself the reason we have such large brains. Nor is it the reason that language evolved in our lineage. Rather, the Scheherazade Effect is more a by-product of the fact that we have large brains and language. It is something we can only do because we have evolved both brains large enough to figure out the witty remarks and the language with which to express them. If anything, our large brains evolved to help us to keep a much larger number of relationships going, to allow us to juggle, as it were, more balls (aka relationships) in our minds. This is because our brains and our language capacity evolved to manage whole social networks, not just individual one-off relationships.

The mystery in the mind

Although most mammals and birds show a strong monogamous brain effect, primates do not. They exhibit a rather different pattern: a very strong relationship between total group size and brain size. It is still possible to detect a monogamy signature beneath this (monogamous species have larger brains than polygamous and promiscuous ones once we partial out the otherwise over-riding effect of group size), but the really strong effect in primates (unlike other mammals) is the global relationship between group size and brain size. What primates seem to have done is to take the psychological machinery that underpins pairbonds and generalise it to other members of the group, so creating friendships. Since the number of friends you can have is limited only by the size of the group, the net outcome that we observe is the relationship between group size and brain size.

To explore this further, Penny Lewis and I undertook a series of studies using brain scanning to explore the relationship between brain size and social network size in individual humans. In these studies, some forty-odd people subjected themselves to having their brains scanned and then completed a list of all the people they had contacted within the past week. It turned out that there was a significant relationship between the number of people you know and the size of the bit of your brain just above the eyes. What is important about this bit of the brain is that it is also implicated in what has come to be known as theory of mind. Theory of mind is the capacity to understand what another individual is thinking – in effect, to have a theory about the existence of other minds. It allows us to state that *I believe that you think something is the case.* Children are not born with this capacity, but acquire it at about the age of five.

Formal theory of mind of this kind is quite a limited capacity. After all, five-year-olds can master it with relative aplomb, and by the age of six children are faultless, skilled practitioners. Adults can do much better than that. In fact, our research suggests that adults can cope with up to four other people's mind states at the same time, plus their own of course – five mind states in all. We can manage that in terms of five different people's mind states or a recursive series of exchanges between just two of us – *I suppose that you think that I wonder whether you want me to believe that* [something is the case]. It's not clear what that really means for us – or indeed why we need to be able to do this at all – but at the very least it suggests a scale of social mentalising competence, a capacity to think through how others see the world, what they believe, what they intend to do, how they might react under different circumstances.

In young children, acquiring the first stage (formal theory of mind) marks a major change in their mental competence. With the acquisition of theory of mind, they can, for the first time, do two key things they had not previously been able to do, and which no other species of animal can do: engage in pretend play and lie convincingly. The second of these is particularly crucial: they can now understand the mind behind your behaviour, and so appreciate much better how they can manipulate your knowledge of the world to mislead you. Before that, they can certainly deceive you, but mainly by virtue of having learned that if they say things in a particularly convincing way, you will likely believe them; they don't really understand why it works, *why* you believe what they say. At this stage, they are good behaviourists – they can very quickly recognise and learn correlations in the way the world works, but they don't understand the causality that underlies those correlations. Mental-state causation seems to be even more difficult to master than simple causation in the everyday physical world, perhaps because we can only infer these kinds of causes by analogy with our own minds, and it takes us a while to learn how to reflect on our own mind states. It takes children considerable time and effort (four years or so, assuming they don't become seriously conscious humans until about a year old) to work their way up through a series of stages in which they first understand causality in the physical world, then use that to extrapolate to their own mental world of beliefs and desires, and then use that in turn to extrapolate to other people's mental worlds. When all is said and done, that is a pretty impressive achievement, one that, with the possible (but still disputed) exception of the great apes, appears to be unique to humans.

But once children have mastered theory of mind, it seemingly allows them to scaffold successive stages so that they can extend the sequence of mind states beyond just one other person and their own mind. By the time they reach mid-teenage, their mentalising skills have advanced sufficiently to allow them to handle four other people's mind states at the same time – which, including their own belief state, means they can manage the five orders of mentalising that characterise normal adults. However, as with all things, there is considerable variation between individuals in these competences. Although the average is five, people vary from around three or four at the low end to six or even occasionally seven at the high end. These differences are probably in part correlated with differences in basic cognitive abilities like intelligence (or, at least, the cognitive processes that underpin intelligence, like the ability to reason causally, to remember facts and relate them to each other, to reason by analogy and so on) and in part to differences in language ability. Language seems to be essential to our ability to reason correctly at the higher orders of mentalising, not because language *creates* these higher orders but because the grammatical structure of language helps us to manage and keep track of the convoluted causal sequences involved.

My colleagues Penny Lewis, Neil Roberts, Joanne Powell and I have been able to show that individual differences in these mentalising skills in normal adults correlate with the volume of neural matter in the core areas that always seem to feature in neuroimaging studies of theory of mind. The areas concerned lie at the two ends of the temporal lobe (the sausage-shaped lobe that juts forward out of the back end of the brain and lies just beneath each ear) and in the frontal lobes just above the eyes. What is significant about our findings is that we were able to show that there is a quantitative relationship between the size of these areas and individuals' mentalising competences – how well they could understand others' belief states. Those who can manage sixth-order mentalising tasks (*I suppose that you think that I wonder whether you want me to believe that you would like* [something]) have more neural matter in these core regions of the brain than those who can only manage four orders (*I suppose that you think that I wonder whether you want* [something]).

More importantly still, we were able to show that these both also correlated with the size of the individuals' social circles. In particular, the two behavioural measures (mentalising ability and size of social circle) both correlated at the same time with the size of one particular part of the frontal lobe, the orbitofrontal cortex, just above and behind the eyes. Since the brain has evolved from back to front, it is these areas above the eyes that have expanded out of all proportion during the course of human evolution as our brain size has increased, thus allowing us to hold together larger and larger social groups. It is this dramatic expansion of the front end of our brain that is responsible for our high foreheads compared to the great apes, or, if it comes to that, compared to any of our fossil predecessors (including our famed Neanderthal cousins). We need the high foreheads to accommodate our particularly big frontal lobes – the very bit of the brain whose loss caused Phineas Gage so many problems after his accident.

In 2004, Andreas Bartels and Semir Zeki scanned the brains of seventeen people who declared themselves to be deeply in love while they were looking at a picture of their beloved and, as a control, pictures of three same-sex friends of similar age. The pictures of the friends provide the baseline here for viewing of pictures of very familiar people, thus allowing us to see which brain regions are specifically active when looking at the picture of a romantic partner (and only when looking at pictures of these individuals). In other words, it allows us to screen out all the other bits of the brain that are involved in looking at any old picture or at the face of any old friend. Bartels and Zeki found there was increased activity in a small number of areas that are involved in various ways with emotions and emotion evaluation (parts of the striatum in the midbrain, the insula and the parts of the cingulate area in the cortex) when looking at the romantic partner, while some areas in the prefrontal cortex, the temporal cortex and around the amygdala show *reduced* activity.

The striatum in the midbrain is commonly activated in rewarding contexts (responses to food and drink reward, as well as to monetary reward), and has been associated with the release of dopamine (which, as we saw in the last chapter, is associated with reward and with positive feelings experienced when viewing photos of those you love). The suppression of response in and around the amygdala is interesting because this region seems to be associated with fear, sadness and aggression. So it looks like viewing pictures of your romantic partner reduces the sense of negative emotions that you might just happen to be feeling at the time and enhances positive emotions. Interestingly, imaging studies of sexual arousal yielded activations in areas close to the cingulate cortex, the insula and the caudate nucleus, all of which were activated by a sense of romantic love, suggesting that the two may not be all that different in terms of their neural wiring – as we might suspect given the ease with which romantic love can trigger rampant sex. The reduced activity in the temporal lobe and the prefrontal cortex is particularly interesting because, as we saw earlier, these are areas associated with mentalising and, in the case of the frontal cortex, with rational thought. So this particular result rather suggests the suspension of active critical thinking, as well as the suspension of mentalising about the individual concerned, and a shift of emphasis towards a purely emotional response.

In a later study, they looked at maternal love, using a group of mothers viewing photos of their own and other people's babies. Once again, the insula and the cingulate cortex were more active when viewing one's own baby, as well as several other subcortical areas in the striatum that were active in romantic contexts, and there was the same pattern of reduced activity in areas like the dorsal prefrontal cortex and the temporal lobe that were deactivated in the romantic context. However, this time they also found activation in the orbitofrontal and lateral prefrontal cortex (areas associated with reward and mentalising) and some additional subcortical and midbrain areas, including the thalamus, none of which was activated by romantic love. Some of the subcortical areas that were activated are known to have high densities of oxytocin receptors, which, as we saw in Chapter 2, are implicated in mother–infant bonding. These regions are also known to be activated when mothers hear a baby cry (although in this case it doesn't necessarily have to be the mother's own baby). One of these, the periaqueductal gray, is also explicitly associated with the suppression of pain during intense emotional experiences like childbirth and so might equally well involve endorphin receptors. It is particularly interesting that this area in the midbrain has a direct neural link through to the orbitofrontal cortex, with its strong associations with reward as well as with mentalising and social network size.

The key point is that romantic and maternal love seem to have some brain regions in common, but also others that are unique to each. They are not, it seems, one and the same kind of thing at all, but may involve quite different motivations and neural mechanisms. The regions in common seem to involve parts of the brain known to be associated with oxytocin and endorphin receptors, and with reward. Even more interesting is the fact that regions associated with fear and aggression (the amygdala) and those explicitly associated with mentalising and the ability to understand others' emotions, intentions and trustworthiness (the temporal pole, the temporoparietal junction and the medial prefrontal cortex) seem to be suppressed or deactivated. It's as though we abandon any pretence at trying to read too much into the behaviour of those to whom we (desperately?) want to be attached, so we switch the socially smart areas off. It's a case of blind commitment, the heart literally overruling the brain.

The mind is still largely an opaque mystery to us, and although neuroimaging technology has allowed us to gain some extraordinary insights into the working mind in the last decade, the technology is still quite crude and it can be difficult to interpret precisely which bundles of neurons are actually firing. Our understanding of what is going on when we feel particular emotions or ponder the future remains largely unknown territory. The technology is, however, developing extremely fast, and the next decade will surely open up new and exciting vistas. Even so, it will still be very difficult to match up what we see on the computer screen with what the person being scanned is actually feeling. There's a translation problem that will always make it difficult to convert what is, in essence, a wiring diagram into real live 'raw feels' as you experience them. So, while we can certainly pinpoint areas in the brain that seem to play a central role both in our mentalising abilities and in our emotional responses to close relationships, we have no real conception of how these areas produce these raw feels.

*

My focus in these last two chapters has been how the brain deals with romantic relationships in an implicit, automated kind of way, seeming to leave us as victims of neurological forces beyond our control. Of course, this is not strictly true. At one level, we evaluate our prospects in the mating market with a considerable degree of forethought. We consider the pros and cons of a particular romantic prospect in ways that are often surprisingly mercenary. Indeed, they seem especially mercenary given both the seemingly spontaneous, unconscious way in which we seem to fall in love and the higher motives of which poets have made such great play. In the next two chapters, I want to dig beneath the surface a bit more to explore the extraordinarily calculating way in which we often seem to approach our relationships.

4

Through a Glass Darkly

Had we never lov'd sae kindly,
Had we never lov'd sae blindly,
Never met – or never parted,
We had ne'er been broken-hearted.
                              'Ae Fond Kiss'

In *Pride and Prejudice*, Elizabeth plays cat and mouse with the deeply eligible Fitzwilliam Darcy, unable to decide whether or not to say 'yes'. She muffs her lines, overdoes it, jumps to conclusions when she shouldn't, but finally triumphs, though not before nearly shooting herself in the foot by dismissively rejecting his offer of marriage. Whatever it is that prevails on Elizabeth eventually to change her mind and accept Mr Darcy's advances, hovering in the background is the single most important thing that makes him a desirable catch for everyone (aside, obviously, from Colin Firth's clean-cut features in the 1995 TV drama) – the fact that he had inherited the enormous Pemberley estate in Derbyshire. In Jane Austen's times, marrying into the landed gentry was the perennial objective of every county girl. Settling for second best by marrying the curate was the sad fate that befell the Plain Janes who failed to catch their Darcys.

Austen's observations of our foibles rank her among the most acute observers of human psychology. Her deep intuitive understanding of the motives that underpin our behaviour are, in many ways, second to none. She had an extraordinary feel for the things that we worry about deep down but are rarely willing to say out loud. Part of the problem is that we are often deeply cynical, motivated by concerns that are banal and nakedly self-interested, but unwilling to admit it in public. This is particularly problematic when it comes to romantic relationships, because if we were really open about our motives it would destroy the magic and undermine the very basis of the relationship that we are intent on creating. We would invite suspicions of gold-digging and opportunism, and prospective partners would be wise to be suspicious. It might even make them want to strike very hard-nosed bargains of their own. We do best to cover up our motives by a thick layer of emotion and a strong dose of Freudian suppression.

Despite this, underneath, most of us do covertly evaluate our options and test the waters before casting our lot irrevocably on the biggest decision we will ever make. And in this chapter, I'll explore just what calculations do lie beneath the anguished, emotional public face we present to the outside world.

Rational choice?

Jane Austen's acute observations of everyday life in the opening decades of the nineteenth century continue to make compelling reading two centuries after they were written. The timelessness of her novels lies in their storylines: how and why we choose the life partners we do. She noted what she saw going on around her with the eye of a trained observer. What she saw is there in the marriage registers of the day. The demographer Eckart Voland has analysed in great detail the historical parish registers of the Krummhörn, the area on Germany's Atlantic coast just above the Netherlands. During the eighteenth and nineteenth centuries, while Austen was observing English county society, girls from the Krummhörn who married into a higher social class did so several years younger than their sisters who married within the class they had been born into. And, conversely, wealthy husbands married women who were much younger than themselves, compared to men from less wealthy classes: they could pick and choose, whereas the less well off had to wait for the ones who had given up all hope of finding their Mr Darcys. The women who eventually married into their own natal class seem to have been holding off as long as they dared in the hope of landing a desirable catch from the social class above, but the risk of being left on the shelf eventually drove them to settle for what they could find at their own social level. But by then they had lost several years in the biological race to reproduce.

With the slow reproductive rate that humans naturally have, even contributing one less child than the average to the next generation is a significant evolutionary cost: it represents as much as 20 per cent of the average number of five surviving children that are born to a woman during her lifetime in natural fertility populations.* In evolutionary terms, that's a massive cost to bear. Moreover, since the risks their children ran of catching diseases and even dying were dictated by social class and its wealth implica-

* That is, small-scale traditional societies. These should not be confused with Victorian society where the benefits of modern medicine radically reduced infant mortality and allowed families of ten or twelve. Nor should they be confused with modern societies that have gone through the demographic transition and reduced average family size to two.

tions (as in fact they still are, even in twenty-first-century Britain), the best-of-the-bad-job option just gets worse the longer you hold off from accepting the hand of your class equivalent of the curate. Really, you would have been better advised to recognise your limitations right at the start and settle for the curate straight away. The problem, of course, is that hope springs eternal and the competition for Mr Darcy is frequency-dependent: it all depends on who else is in the competition. You might just be the lucky one on whom his wandering eye happens to fall.

This may seem cynical, but it is a reflection of the harsh reality of life at the time. Being left a spinster was a tough option in the eighteenth and nineteenth centuries, and often meant a life of servitude and drudgery in one's parents' household or that of a married sibling. You can see the same decision process in the life histories of girls who became pregnant out of wedlock. Typically, young women at all social levels in the nineteenth century married slightly older men, as they still largely do. But girls with illegitimate offspring very often married men who were younger than themselves, a sure sign that they were settling for second best – a man with few resources to offer, someone who lacked prospects, but better than being left to struggle on the scrapheap with nothing at all.

Courtship itself is a protracted process of negotiation, as Austen recognised. It's a game of poker in which we don't know for sure the intentions of the players around us, so we make tentative bids to smoke out the serious options. We watch the signals that come back carefully to assess the likely response. Was that an accidental brush of the hand? Or was it deliberate? Once, in medieval times, ladies dropped their handkerchiefs near the young knights who attracted their interest. It made everything so much easier, especially for the socially less sophisticated sex. Alas, these niceties of the wimpled age no longer apply, and the game is just that much harder. But, as in poker itself, it is the eyes that give the game away. Once our interest is roused, we just cannot keep our eyes off the prize.

In fact, that process of evaluation and indecision plays an important role, especially for romantic relationships. These are not like friendships that we can pick up or drop at a moment's notice, relationships that we can slowly allow to wither away by simply doing nothing. Once we have switched on the emotional turbocharge, there is a great deal more at stake. Breaking out of the relationship later can only be done at great emotional cost, often to both parties. So it's desirable to get it right first time. And that's why courtship is a series of phases of increasing intimacy: at each stage, you go so far, and then pause to evaluate before deciding whether to go to the next level – or head for the exit while you still can without creating too much grief.

But what constitutes the ideal partner? One invaluable source of information on what people want in a partner that we have explored in some detail is what they say in those lonely hearts advertisements. Here, in a nutshell, is what the punters think they have to offer and what they yearn for in the ideal partner. So let's see what we can learn from these literary gems.

How lonely the heart

Personal advertisements provide a succinct vignette of what people think is most important to them in relationships. Here, in black and white, is the distilled wisdom of their experience in the sandpit of life. Because you cannot write an essay but have to sum up what you want to say in a dozen words, a certain discipline is imposed on the writer, focusing the mind on the handful of criteria that they think are most critical. What do *you* think is most important in a prospective partner, and what do you think *they* will be most interested in about you? Amazingly perhaps, the answers to both questions seem to boil down to a very small number of traits, although we can use a variety of words in describing ourselves within each of those traits. The essential criteria are just six in number: age, physical attractiveness, status and wealth (two sides of the same coin, really), social skills, hobbies and interests (which might reflect intelligence), and commitment. Sometimes the wording can be extraordinarily subtle, even completely opaque to those who don't know the code – a casual mention of an upmarket London postcode to indicate someone who is well-heeled, a literary allusion to indicate someone who is well-read and cultured. Occasionally, they can be bluntly explicit – though, in my experience, mostly from males seeking nothing more than casual extramarital sex.

In our detailed analyses of the advertisements of the two sexes, one thing is quite striking: male and female advertisements tend to be mirror images. Men tend to emphasise that they are seeking the first two of the traits I listed above (age and attractiveness) and advertise the other four, whereas women tend to advertise the first two and seek the other four. This is a reflection of the rather basic biological fact that female mammals have more to lose by a mating than males do: once they have conceived, female mammals are unavoidably committed to a lengthy and expensive period of gestation and an even longer and more expensive period of lactation. To launch oneself down that road, and then pull out before the end is to waste time, energy and resources, not to mention the emotional cost. Not surprisingly given this, women tend to be more choosy than men, or at least to be more choosy during that period of their lives when it really matters. Here are two typical ads, one male and one female, to illustrate the point:

> Good-looking male, 36, professional, solvent and ambitious, likes travel, keeping fit and theatre, seeks similar female.

> Divorced white female, 42, attractive, slim, fit, would like to meet single/divorced white male, 38–48, handsome, professional, home-owner, non-smoker, college-educated, financially stable, good values, no drugs or drink.

He is rather unspecific about what he is looking for, but knows that cues of status and wealth and of intellectual hobbies will go down well with the punters; she mentions the only two things that men are ever interested in and then launches into a long list of criteria the lucky guy must satisfy if she is to give him even so much as a second glance.

Although the balance between the traits does vary according to local circumstances, the broad picture is exactly the same around the world, in different cultures, and historically. In her study of late Victorian marriage advertisements, Helen Pearce found very much the same pattern of sex differences. And in a large-scale questionnaire study of desirable traits in a mate involving over sixteen thousand people from fifty-two countries on all six continents, the same broad patterns emerged. There was also a large and cross-culturally consistent difference in the desired number of sexual partners between the two sexes. Typically, men envisaged having around six partners in a lifetime, but women only envisaged two. Broadly speaking, the men were more interested in sexual variety and having more short-term relationships than the women were. There is, of course, some variation both between individuals and, within individuals, across the lifetime. In the peak reproductive years, for example, women tend to list more traits that they want in a partner than men do, and to offer many fewer. Later, as they approach and go beyond menopause, they will become less choosy and lower their standards.

While a female mammal finds herself committed to a long period of investment once she conceives, a male mammal can always walk away from a mating and find another available female. In effect, a male mammal makes a judgement about how he can maximise the number of offspring he produces over a lifetime. Staying with one female has advantages in that mating access is guaranteed and can be defended, but in the end the numbers of offspring are limited by the slow rate at which she can produce them. For the male, there is a lot of idle time in between. Roving is more risky, but if the male can search for, find and mate with enough receptive females, he may do better than his more faithful alter ego. The issue hinges on how effectively he can find receptive females. If they are hard to find, he will do best to stick with one once he has found her; if they are easy to find, he'll do best to go roving.

So it is perhaps not too surprising that, in general, female mammals are notoriously more 'coy' than the males and make more careful choices. We do not often find females strutting their stuff on mating arenas in the way that peacocks and many male antelope do – displaying their physical wares to allow the watching females to choose the best among them. These effects are exacerbated in the human case by the more complex nature of parental investment. Unlike most deer and antelope, whose offspring are pretty much independent and off their parents' hands as soon as they are born, the offspring of monkeys, apes and humans require continuous investment over a very long period of time, first in terms of milk and then in terms of socialisation and protection. We humans continue to invest in our offspring at least until they are safely placed in the adult world – twenty years later . . . thirty years later . . . perhaps even forever? That makes it all the more necessary to be sure that the deal will work, since being left holding the baby even as late as the second or third decade of the child's life can still affect its opportunities. Single-parent families appear disproportionately in the poorest sector of the population in all societies, even in modern welfare-state Britain.

In the raw state of nature, there are two very different sources of paternal input that women might be interested in. One is the quality of the male's genes and the other is his commitment to the pairbond, together with the products of his foraging and any socialisation investment in the offspring that he might be prepared to make. Once heritable wealth came into play, a third potentially important dimension was added: the wealth that the male or his family had to offer. Wealth is important to offspring survival and rearing because it allows the mother to produce or acquire more food or other resources to invest in her offspring. Many studies of both contemporary pastoralist and horticultural societies, as well as historical agricultural societies such as the Krummhörn, show that women married to wealthier men (whether this is measured in terms of camels or land) rear more surviving children than women married to poorer men, even when the absolute numbers of offspring born are held constant. Wealth is still an important factor in offspring illness and physical growth rates, as well as educational achievement, even in modern industrial societies like ours: those from lower down the socioeconomic scale have smaller offspring on average, and their children suffer higher illness and mortality rates as well as achieving less (mainly because the parents can't afford to invest so heavily in their education, health or life opportunities).

For women, then, there are at least three different dimensions they can try to balance, and they won't necessarily all coincide. The men with the best genes may not be the wealthiest, and the wealthiest may not be the most committed. In the end, women's choices end up being a compromise, the best of a bad job. How they weigh the various factors may vary somewhat from one set of economic and social circumstances to another, but broadly speaking all of them appear in the equation and have to be balanced against each other. In contrast, men largely seem to be driven by the dictates of a single interest, namely fertility – at least, judging by their personal ads. As a result, male mate-choice preferences are relatively simple, and in principle easy to satisfy. In contrast, female preferences are more complex and can probably never be fully satisfied – they are always making do.

In humans, this difference in what we might call demandingness between the sexes in part reflects the weightings that the two sexes give the six criteria. No matter how old men are, they tend to advertise for women of about the same age – in Western cultures, typically ones in their late twenties, when they are at the peak of their fertility. So much so, in fact, that in our samples even men aged between eighteen and twenty-five asked preferentially for women in their later twenties, as did the forty-to-sixty-year-olds. This preference for women in the later twenties is not, of course, universal. In some cultures, and especially those where child brides are the norm, men prefer women of much younger ages, even teenagers. This is probably because men adjust their age preferences to coincide with the normal age at which women first reproduce in the population where they happen to live.

My Polish collaborator Boguslaw Pawłowski and I showed that, for the UK, men's preferences for women of different ages matched women's natural fertility pattern in the UK (i.e. the probability that women of any given age in the UK would give birth) almost perfectly. And better still, it matched women's reproductive potential – in other words, the number of future offspring that the average woman of a given age can expect to produce in the remainder of her lifetime. It may well be that this is a reflection of socially enforced monogamy. Marrying a girl as soon as she is likely to reproduce maximises the number of offspring you can expect to sire, given that you are then stuck with her for the duration. Marry too early, and you may lose out on any additional offspring that you might have gained through casual relationships before being forced to withdraw from the game and become monogamous (or, at least, before casual relationships become socially and legally more difficult to pursue). But when adultery is strictly punished, as it is in many Islamic countries in Africa and the Middle East, it's particularly desirable not to waste time and opportunity, so the younger the bride the better – which will tend to drive the average age at which women get married (and reproduce) down towards its natural limit at puberty. In more traditional societies, such as those of hunter-gatherers or even our own, that operate a mating system that is closer to serial polygyny, age at first marriage is less critical because each 'marriage' only lasts long enough to fit in one or two offspring, and then you are on to the next. However, there will still be a general pressure in favour of women who are at their peak of fertility. Exactly what age that is will depend on whether contraception is available to delay the age at which women typically have their first baby. One thing this does seem to imply is that women dictate their own fertility, and men adjust their preferences in the light of this.

All else being equal, however, a woman's fertility is, for better or worse, a simple function of her age, increasing from puberty to a peak in the twenties, and then declining slowly across the next decade or so before precipitously dropping away to zero with the menopause around the mid-forties. Age is thus a reliable cue of a woman's fertility. Fertility is, of course, ultimately a function of the levels of oestrogen floating around in the body, and oestrogen has a strong influence on physical appearance, creating the smooth shiny skin, glossy hair and hourglass-shaped, slightly plump figure that we associate with women in their twenties. Male reproductive biology is not so constrained functionally: even though libido may decline with age, the capacity to sire children remains more or less constant. Age-wise, almost any male will do, and that leaves his other traits to become more important from the female point of view. The net effect seems to be that women prefer men who are just a few years older than they are.

The more astute will have noticed that, before we even get any further, we are already into a battle of the sexes. One sex wants a much younger wife, and the other wants a very slightly older husband. They can't both win, except for a brief period around the later twenties when their respective preferences happen to coincide. After that, their preferences diverge progressively. Evolutionary theory would tell us that, because women have more at stake than men, they should be more demanding and exert more pressure to get their way. And that's exactly what we see: married couples, or even those in partnerships, tend to be more evenly matched in age, broadly matching women's preferred age difference of three to five years. The exceptions are *some* second marriages for men, when the woman can be as much as a decade younger (the classic younger second wife), or women married to *very* rich men – the proverbial case of the glamorous young thing married to the ageing rocker, in which case the age gap can be almost as wide as you like. The first of these two exceptions is actually rarer than you might suppose, with most second wives actually being only five to ten years younger than their partners. The second, however, is more interesting because, in one sense, it is very much the exception that proves the rule – precisely because such marital arrangements rarely last long: ageing rockers tend to have a hard time hanging on to their glamorous young things, who, on the other hand, tend to leave the marriage rather pleased with the alimony – or the inheritance, whichever comes first.

The Cinderella moment

The examples we discussed in the previous section point to an important feature of our mate-choice strategies: our capacity – and willingness – to compromise on our ideals when push comes to shove. We do not operate in an ideal world, we operate in a market. We are not free agents able to take our pick of the best fruit on the supermarket shelf, in the way that some animal species seem to do. In a species like the peacock, the males all line up on a mating arena or lek, each with his tiny mating territory just a few tens of yards in diameter. There, they sit like shopkeepers in the souk, waiting for passing customers to come their way. Should a female appear, all the males are up and strutting their stuff, tails erected, feathers shimmering. The females inspect the wares on offer, and, should they think one a half-decent deal, will stay on his territory and mate. Then they will go away to lay their eggs, fertilised by the male's sperm, in the nest they have built for the purpose far away from the lek. The males take no further part in the business of reproduction, but return to their waiting game for the next female. For males and females alike, it's essentially a one-armed bandit.

Humans, however, play at a two-armed bandit. Like all species that form longer-lasting pairbonds, we may well have preferences and ideals, but then so does everyone else out there, and that means we have rivals and competitors who may beat us to the prize. We face a real prospect of having to settle for second best. We risk going home from the ball alone. Or we might have to settle for the ugly one that no one else wants. But it is also in part a consequence of the fact that human mate-choice strategies are complex and multi-dimensional compared to those of most other mammals and birds. That alone forces us into making compromises even in the absence of rivals, because, sadly, it is almost inevitable that no one individual will be perfect on every single dimension. Because women and men have rivals for their respective ideal mates, we are inevitably forced to compromise at two levels (between the various dimensions of the ideal partner and between rivals). If someone else gets the perfect husband first, you are forced to settle for second best, so you'd better have some principled algorithm for deciding how to compromise on your ideals, for prioritising your criteria. Economists use the term *satisficing* for this: you're trying to satisfy as many different criteria as possible, and inevitably you can't do it perfectly.

Even though wealth and status have a very substantial impact on the effectiveness with which women can rear their children in traditional (pre-industrial) societies, they are always embroiled in a three-way tussle with gene quality and commitment. Nonetheless, wealth and status will always have a bit more of an edge – especially in the impoverished socioeconomic conditions in which most people find themselves – simply because of the effect wealth has on offspring survival as well as the offspring's future social and reproductive opportunities. This is reflected rather clearly in what has become known in evolutionary ecology as the polygyny threshold. In essence, it marks the point on the male wealth scale where it pays a woman to become the second wife of a polygamous male rather than be the sole wife of a poorer monogamous male. If the male earns or owns enough for two wives to share so that they still do better than they would if they married any other male monogamously, then there will be pressure for polygamy. What drives this is a steep inequality of wealth across males: some males have to be *very* much richer (aka ageing rockers) than the average for it to work. Where wealth differentials are low, there is less pressure for polygyny. Polygyny may not be ideal – and there are costs to polygamy for women – but nothing in real life or evolution is ever ideal. The point is that it's a trade-off, and it's one that the women ultimately adjudicate. It isn't just imposed by the patriarchal hierarchy as some like to suppose, but rather is a consequence of what women will allow. Men will exploit whatever loopholes they are allowed, but if it isn't in the women's interests, there is not a lot the males can do about it. You just end up being the peacock that no one will visit.

The problem for most of us blokes is that wealth and status do not usually come early in our lives. The few for whom it does, like young Darcy, are invariably snapped up very fast. For the rest of us, it's a slow process of accumulation. The real problem for women is that as men age, they become increasingly at risk of dying. And given that we have to invest in our children for at least twenty years to set them on their way in life, that's a bit of a problem. We have been able to show, using UK personal ads, that women adjust their preference for men of different ages as a direct function of the trade-off between the men's natural accumulation of wealth with age and their age-specific risk of dying or abandoning their wives (with these three variables estimated directly from the UK national statistics). The most desirable men (relative to their availability in the population) seem to be those in their early forties. After that, it is downhill all the way to the Zimmer frame. As with all trade-offs, there comes a point when a male is so rich it pays to marry him even if he dies prematurely . . . although, come to think of it, perhaps that's an advantage? In other words, the exception that proves the rule is this: if you are going to go for an old one, just make sure he's a really super-rich old one so that if he dies on you, you come out of it smiling. It's the ageing rocker syndrome.

The world of lonely hearts ads, as with everyday courtship, is full of such trade-offs. The fact is that we may aspire to something, but when it comes down to the wire we will, like Jane Austen's ageing spinsters, accept whatever we can get rather than go without. For example, in the world of personal ads, men won't even bother to reply to advertisements by older women – and that's women older than, say, their late thirties. But the realpolitik of life is that if the women can persuade them at least to meet up, they will likely compromise on their ideals rather than walk away empty-handed. It's a bird-in-the-hand-is-worth-several-in-a-lonely-hearts-column issue. So the problem for older women is just to stay in the game and not have their ads binned before they have had a chance to meet a bloke. To do this, they resort to subterfuge.

When Boguslaw Pawłowski and I looked at UK personal ads, we were struck by the fact that a surprisingly high proportion of women didn't say how old they were – especially given how important age was as a criterion for mate choice in men's minds. After a while, we began to suspect that it was only older women who were being coy about their age. Younger ones almost always specified their age. But how to figure out the ages of the women who didn't declare it? Then we had an idea. Because women typically ask for men who are three to five years older than themselves, we figured that by using the age of the males such women were seeking, we could estimate how old they actually were. It turned out that while only 4 per cent of advertisers in their thirties declined to give their age, 21 per cent of those in their fifties failed to mention theirs. Moreover, older women who declared their age were more likely to give a decadal age rather than an exact age (stating their age as 'fifties' rather than, say, fifty-seven). In other words, it was mainly women at the end of their active reproductive life who were suppressing this information. The reason was not hard to see: women who did not give their age were able to be much more demanding in their ads than women of the same age range who made the mistake of declaring their age. In fact, women who suppressed their age were about as demanding as twenty-five-year-olds, despite being as much as two decades older. Concealing their age allowed these women to be more choosy without putting prospective punters off. That way, they could maintain more control over the situation and be in a position to choose rather than having their options foreclosed by someone else.

Women are not the only ones to do this, of course.

We also noticed that heights over six foot were frequently mentioned in men's ads, but never heights below that. Shorter men just didn't declare their height – as it turns out, for the very good reason that their ads were also discriminated against. I'll have more to say about this in the next chapter.

Having children from a previous relationship can also be a problem, at least for women. Many years ago, I was told by one of the well-known dating agencies that they always advised women not to mention children if they had them. Inevitably, the women ignored this well-intentioned advice. Because their children are invariably the centre of their lives and much to be justifiably proud of, women want to mention them in their ads: their children are part of who they are as a person. It was only when the women came back in tears because they had received no replies *at all* to their adverts that the agency was able to persuade them that they should leave the children out. And then all would be sweetness and light, and the replies came in. This is not a new problem. In the Krummhörn during the eighteenth and nineteenth centuries, a young widow with one child would have a dramatically increased chance of re-marrying if her child died. However, those with more than one child usually settled for widowhood and put their efforts into rearing the children they had. For them, there was less biological pressure to have more children. John Lycett and I demonstrated a similar effect in modern British women, though in this case it related to the rate of abortion. Using the UK national statistics, we were able to show that the probability of an unmarried pregnant woman having an abortion was directly related to her age-specific expectation of getting married in the future and so having legitimate offspring. The closer she was to menopause and the less likely to have more offspring within a future relationship, the more likely she was to continue with the pregnancy.

The main reason for this, biologically speaking, is very simple: by and large, males don't want to have to raise other men's children if they can help it. Doing so means that they have less to invest in their own children, and, from a biological viewpoint, such altruism is evolutionarily counterproductive. But if their options are limited, then even males will compromise rather than go away completely empty-handed. David Waynforth and I found this rather clearly in an analysis of American personal ads. Males who indicated that they had resources often explicitly stated 'no children from previous relationships'. But males who lacked resources (or at least didn't advertise having any) were more accommodating and would sometimes indicate their willingness to take on children from a previous relationship. The anthropologist Barry Hewlett noted a similar effect among the Aka Pygmies he studied in the Congo: males who were regarded as good hunters (an attractive trait to Pygmy ladies) invested little time and effort in childcare and preferred to use their time to pursue liaisons with other women. But the poorer hunters with less to offer tried to compensate by greater willingness to help out with the children.

The point of all these examples is to remind us that the mating game is actually very complex. We might have clear preferences, but the world we live in is imperfect and we can rarely exercise those preferences to the full. Most of the time, we are forced to compromise. Our experience of life allows us to fine-tune how much we are willing to compromise with surprising accuracy. When Boguslaw Pawlowski and I analysed UK lonely hearts ads, we found that, across age cohorts, both men and women adjusted their demandingness (how many traits they specified for a partner) according to their perception of their own standing in the marketplace of love. Individuals of the age category that was most sought-after relative to their availability in the population (later twenties in women, early forties in men) were most demanding of prospective partners, but as their market value declined, so they increasingly compromised on how demanding they were. Both sexes were surprisingly astute in matching their demandingness to their market value – with one exception: men in their later forties radically overestimated their hand. Still, seemingly even men can learn, because by the time their reached their early fifties, they were back where they should have been.

Humans are very sensitive to the costs and benefits of the options available to them. Emily Stone and her colleagues surveyed the mating preferences of some 4,500 men and 5,300 women from thirty-six cultures round the world. They were able to show that men dropped their standards of mate choice as the sex ratio became increasingly male-biased (so creating more competition for them) and raised their standards, at least for long-term mates, when the sex ratio became more female-biased (giving them more choice). More importantly, it seemed that men switched to more casual sex when they were in the minority (so that women were forced to compete for men and, as a result, could exert less power over them). When there were fewer women available and men were forced to compete for women, men became more willing to accept committed relationships. Women, however, showed the converse effect, being more choosy when they were in the majority. This seems to be a response to the fact that, under these circumstances, there are more men seeking casual relationships because they can get away with it, so women try to exert pressure on them by being more demanding. This suggests that the interaction between the two sexes' strategies is dynamic and may never achieve a stable equilibrium. It also suggests, once again, that wars of the sexes easily arise when men's and women's strategic interests conflict.

My favourite example of our capacity to compromise was a study carried out in the singles bars of Baltimore, Maryland, by James Pennebaker. Sober punters of both sexes were asked to rate members of the opposite sex and their own sex at three times during the evening (9 p.m., 10 p.m. and midnight). While there was no difference in the rated attractiveness of members of the rater's own sex (if anything, they got less attractive as the night wore on), members of the opposite sex mysteriously became *more* attractive as it got closer to the Cinderella hour. (There was, incidentally, no difference between the sexes.) Now, unless you are prepared to believe that the ugliest people scored first and left the party early with Prince Charming, it seems to me that this is sure evidence that punters were dropping their standards as the prospect of going home alone loomed. Who said romance was dead?

Showing off

Conventional wisdom has always assumed that males go out hunting to bring home meat for their spouses and offspring. Given that, it seems only reasonable that women might want to select men who are good breadwinners or have wealth and resources to offer. But doubts about the significance of hunting as a form of parental investment began to surface in the 1990s. Yes, men do go hunting, and they do bring meat back to camp. But maybe the *point* of it isn't the meat itself? When the American anthropologist Kristen Hawkes looked more closely at hunter-gatherer societies, she came to the conclusion that big-game hunting only has a modest impact on food provision for the family, given the effort invested in it. To be sure, when a man gets an elephant, everyone gets a lot of meat. But that only happens at very irregular intervals, and males have to invest a great deal of time, effort and risk tracking and stalking elephant (or any other big game) to get one. In reality, men would contribute many more calories to the family by trapping small prey or by gathering vegetable foods with the women than they did by hunting large dangerous animals – *and* they would expend a great deal less time, energy and effort and run less risk to life and limb doing so. But there was one further puzzle in hunter-gatherer societies: when men do bring home meat, they don't give it all to their families. Strict rules prescribe that they have to share it with everyone in the camp, and their own family often ends up with a very meagre amount. Hawkes began to wonder whether the whole point of big-game hunting in these societies might not be the very fact that the prey is large and dangerous. Being able to prove that you could take on these odds and come home with the bacon was incontrovertible proof of your agility, strength, stamina, cunning – in short, the quality of your genes. Hunting big game was a way of showing off, not a way of providing food for your family.

When Darwin was developing his theory of evolution in the second half of the nineteenth century, he struggled with cases like the peacock's tail. How could such an ungainly ornament that so obviously impeded the poor bird's ability to fly have evolved by natural selection? The answer, he eventually realised, was that natural selection acted on an individual's ability to leave descendants, not just its ability to survive. The real issue is the trade-off between survival and reproduction: just surviving for a long time was not of itself enough to guarantee leaving any descendants, which is the engine that drives natural selection. You need to do some breeding as well. And the pre-eminent importance of reproduction implies that you could in principle maximise the number of descendants you leave by breeding fast and dying young. The peacock's tail, Darwin began to realise, could be explained by the power of what he came to call sexual selection, selection for traits that maximise the capacity to reproduce because they make you attractive to the opposite sex.

Sexual selection has two main motors: female choice (females choosing between males who offer different qualities) and male–male competition (males fighting it out to gain control over matings with largely passive females). The peacock's tail was an example of the first process. And an important element in that, we came to realise only a century after Darwin had died, was that traits that honestly signal quality will tend to become the favoured basis for females to choose between males. Now known as Zahavi's Handicap Principle, after the Israeli ornithologist Amotz Zahavi, the central point is that a trait which handicaps the holder is indisputable evidence for the quality of its genes. In effect, the peacock is saying to the peahens: just look at me, my genes are *so* good that I can afford to handicap myself with this enormous train and still escape predators. Darwin's insight had been to identify one of the most powerful of the forces of natural selection, one capable of exerting rapid and often dramatic change in a species' appearance. And showing off lies at the heart of this.

To test the possibility that men might use showing off as a mating tactic, Wendy Iredale, a former student of mine, offered men and women a reward for playing games as part of an experiment, and then at the end of the experiment invited them to consider contributing some, all or none of their earnings to a charity while being watched by either a member of their own sex, a member of the opposite sex or no observer. On average, men and women indicated that they would be willing to contribute between 30 and 40 per cent of their earnings to the charity, with little to choose between the three conditions – except for men being watched by a woman, who volunteered nearly 60 per cent of their earnings. Showing off by being generous was evidently considered a good mating tactic by men, presumably because women find this attractive. Because men seem not to value generosity in women, there is no advantage to women in acting unusually generously when men are watching.

There is more to showing off, however. Sue Kelly, another of my former students, asked women to rate their preferences for seven fictional males whose characters were counterbalanced between heroism, altruism and professionalism. Thus Jim might be a fireman who behaves

both heroically and altruistically and does so profession-
ally, while Fred is a nurse whose behaviour is profes-
sionally altruistic, but not necessarily heroic; George, in
contrast, is a supermarket manager who is not altruistic
but once rescued a colleague from a bank robber (and
whose heroism was thus voluntary rather than professional). Asked to rate vignettes of this kind, women exhibited a striking preference for forming long-term relationships and non-sexual friendships with altruists, but very
strongly favoured heroes (and especially professional heroes) for short-term flings or one-night stands. Being a professional hero sorted the sheep from the goats (or should
it be the goats from the sheep in this case?) in a way that
simply couldn't be faked by bar-room braggarts. These
findings suggest that women may exhibit a preference for
cues of gene quality when selecting whom to mate with,
but prefer nurturing qualities (those that impact on child-
rearing) when choosing whom to live with.

Men are not innocent pawns in this game, of course:
they clearly understand the rules and they use their wits
as best they can to exploit female preferences. In Kelly's
study, men asked to rate the vignettes from a woman's
perspective exhibited almost the same patterns as the women, except that they tended to exaggerate women's relative preferences. It seems as though men read the runes
quite well. So, if risk-taking is for some reason attractive
to women, then taking risks may be as good a way as any
other of demonstrating one's fitness to a prospective mate.
And the bet may be a good one for women: any game
that tests males' qualities and competences – irrespective
of whether it is in the hunting, economic, literary, social or

sporting domain – may provide a perfectly sensible arena in which to sort the men from the boys.

Another example of how risk-taking might be attractive to women involves the Cheyenne American Plains Indians. Historically, the Cheyenne had two kinds of chiefs: peace chiefs and war chiefs. Peace chiefs were largely responsible for the smooth running of community affairs. They did not take part in raiding or inter-community fighting and they inherited their status from their fathers. War chiefs, on the other hand, were responsible for leading the tribe in battle. They were required to forswear marriage and often took an oath never to leave the field alive unless their side won – and to ensure this they sometimes tied themselves to the ground during a battle. Needless to say, many war chiefs died young. However, those who survived did rather well for themselves. After a distinguished career, they could absolve themselves of their oaths and marry. Being pumped full of testosterone (unlike peace chiefs, of course), they invariably had lots of children – and those were just the legitimate ones. When I calculated the lifetime number of children produced by the two kinds of chiefs from the nineteenth-century census records and multiplied these by the relative frequencies of the two kinds of chiefs in the population at the time, it was clear that the two strategies were in evolutionary equilibrium. They were equally good strategies to pursue.

But the important thing about war chiefs was their origin. Almost all of them were orphans, and in Cheyenne society widows and orphans were at the bottom of the social pile, invariably treated rather badly, and certainly lacking in serious marriage and social prospects. For an orphan, becoming a war chief was a lifeline, albeit a risky one that might end in a premature death. But for those physically good enough to make it through to the point of 'retirement', nirvana beckoned. It was a high-risk, high-gain strategy that was not for everyone. The son of a peace chief wouldn't have considered being a war chief because he had a perfectly good route to social and marital success open to him by virtue of inheritance – and one that carried no significant risks. But an orphan only had a choice between a very rough deal at the bottom of society (with little likelihood of marriage and probably an early death anyway) or taking some risks and coming out of it rather well. It was the risks that played the crucial role: you could only afford to go down that route if you were tougher than average, otherwise a *very* early death beckoned – in which case, you would have done better to settle for being a dogsbody that everyone else kicked around. It was an uncompromising test of good genes.

Risk-taking is deeply embedded in the male psyche. Once boys reach teenagehood, their mortality rates suddenly increase dramatically, whereas those of girls do not. Boys race cars, take dares, test themselves in physically demanding and risky situations, play with dangerous weapons, and take drugs . . . and, as a result, sometimes come off badly. This is not to say that girls never do these things, but rather that, as a group, girls are more cautious than boys and tend to take fewer risks. You need look no further than the national mortality statistics. In the USA between 2000 and 2007, for example, the average death rate per year (from all causes of mortality) for white teenage males aged fifteen to nineteen was 65.7 per hundred thousand, whereas for white teenage girls it was just 39.9, about half that for boys of equivalent age. More than half of all male mortality (34.1 deaths per year per hundred thousand) was due to road accidents, and another 15 per cent (10.9 deaths per year per hundred thousand) due to firearms. The equivalent figures for girls were 19.8 and 1.9, respectively, and I'll bet you anything you like that in most of the cases where girls died in road accidents it was actually boys that were driving.

The fact is that boys are just inveterate risk-takers. We showed this in a couple of very simple little studies. Rajinder Atwal, another of my students, recorded how far away oncoming cars were from a pedestrian crossing when men and women started crossing. Men took far more risks than women. More importantly, they were much more likely than the women to cross under risky conditions (defined as an oncoming car less than fifty yards away when the lights were green for the car) *if there was an audience of girls present*. For the girls, the likelihood of crossing under risky conditions was not affected by whether or not there was an audience, or whether the audience consisted of boys or other girls. Boys are just more prone to showing off; and since showing off is best done by taking risks, the more serious the risk the better it sorts the men from the boys.

*

Although this chapter has emphasised the fact that we seem to be quite calculating when it comes to choosing our mates, not everything about mate choice is done with such explicit attention to what we might get out of a relationship. Life is too short for us to be able to check out every detail of a prospective mate's value. Instead, we rely on quick and dirty cues that evolution has honed to provide us with easy-to-spot assessments. Some of these are so subtle, it is always a wonder to me that we can spot the tiny differences we seem to be able to judge in features like the relative lengths of someone's fingers. Life, it seems, is full of subliminal advertising that we respond to but aren't always aware of.

5

Saving Face

Auld Nature swears the lovely dears
Her noblest work she classes, O;
Her 'prentice han' she tried on man,
An' then she made the lasses, O.
    'Green Grow the Rashes, O!'

Some five thousand years ago, Rameses the Great observed of his favourite wife, Nefertari, that her 'buttocks are full but her waist is narrow . . .' A timeless comment on the fact that women tend towards a natural hourglass shape – wide hips and chest, with a narrow waist. The narrow waist is not of itself that exciting, even though Victorian women famously endeavoured to exaggerate theirs by reining them in with corsets (even to the extent of surgically removing the lower set of ribs to produce their famous 'wasp waists'). The real departure from the androgynous body shape of pre-puberty is the enlarged breasts and hips, created mainly by layers of fat deposited in these regions. We use a wide variety of physical cues to gauge the extent to which a prospective mate satisfies our various criteria. Many of these are facial, but by no means all. Body shape plays a role. These cues are often processed automatically and simply kick in a 'yes' or 'no' response.

Something in the way she moves

Humans are unusually fat by primate standards. In normal, healthy, average-weight women, about 20 per cent of body weight is fat, with men closer to 15 per cent, compared to around 3 to 5 per cent in monkeys and apes.* The main reason that we are so fat is to fuel the costs of producing such large-brained infants. The costs of gestation and, especially, lactation are so enormous for us that women need to have massive reserves of fat that they can metabolise to convert into energy for the growing baby over the long haul of parental investment. In natural fertility populations, that amounts to nine months' gestation plus as much as three to four years of lactation. Obviously, the costs are minimal during the first few months of gestation (when the embryo is tiny) and the last six months or so of lactation (when the infant is beginning to feed for itself). But the period in the middle, when everything the baby needs has to come through the mother, is massively expensive.

Under natural conditions, mothers find it hard to eat enough to replace the energy that is being drawn off by the baby, and they undergo considerable weight loss as a result. Without a significant store of fat to draw on, they would be driven below the point of no recovery and start to metabolise their own muscle mass to keep the baby alive. Evolution has seen fit to place these fat depots close to where they are most needed (the breasts and the womb), presumably to minimise the costs of moving energy around the body. The laying down of body fat is mediated by the amount of oestrogen being produced by the ovaries, and for this reason body fat in these key depots may be a good index of a woman's fertility. However, excess body fat in

---

* Monkeys and apes can, of course, become obese in captivity, but only when they are fed an excessively rich diet and allowed too little exercise.

other parts of the body has a negative effect on fertility, thus reinforcing the importance of the slim waist and the hourglass shape.

Because of the need for fat depots to provide energy reserves during pregnancy and lactation, men and women have very different body shapes. The natural male body shape – and that of girls up until puberty – is tubular, whereas after puberty that for women, thanks to the location of these fat depots, is hourglass-shaped. Sexual selection acted on this distinction to make the figure-of-eight shape sensually attractive to men, and in turn on women's bodies to exaggerate their appearance. Conversely, the more tubular shape of males, with their broader, more muscular shoulders, is equivalently attractive to women. Body shape is conventionally indexed by the waist-to-hip ratio (WHR), waist size divided by hip size. In men, a WHR of ≈0.9 is rated by women as the most attractive (in other words, just enough waist to keep your jeans from falling down), whereas men find women with a WHR of ≈0.7 (the shape typical of centrefold models) to be the most attractive. A WHR of 0.7 is roughly equivalent to the more conventional 36-24-36.

The psychologist Dev Singh and his colleagues have obtained similar results from a variety of African, South East Asian and Polynesian cultures, suggesting that these preferences aren't entirely the result of cultural influences from the Western media's hang-ups with centrefold models. However, other studies suggest that, in traditional societies, sheer body mass may be a more common basis for choice. It seems likely that these apparent differences actually reflect the fact that males' preferences for female body shape are fine-tuned to the best predictors of female fertility in the local environment. A plump Rubenesque figure may be a better index of a woman's biological fitness in nutritionally stressed populations (i.e. most traditional small-scale societies), whereas WHR may be a better index in well-nourished populations. One likely reason for this is that a mother's body mass affects how much spare energy she has to put into an infant, and in traditional societies this is probably the single most important determinant of infant growth and survival. In contrast, when survival pressure is markedly reduced, as it is in modern developed economies, the mother's fatness may be much less important for her reproductive prospects than her fertility. In other words, men may switch back and forth between these two key traits as a function of economic circumstances.

By chance, Boguslaw Pawłowski and I turned up some evidence to support this suggestion. We showed that, in a sample of several thousand Polish women, WHR was the better predictor of neonatal weight (a key factor determining infant survival, and hence an index of the mother's fitness) in larger-bodied women (those weighing over eight and a half stone, or about 120 pounds), but body mass index (weight divided by height, a widely used index of absolute fatness) was the better predictor in lighter women (those weighing less than eight and a half stone). Women in most small-scale traditional societies tend to fall at the lighter end of the weight distribution, and this might explain why males in these societies are more influenced by body mass when judging women's attractiveness.

Irrespective of this, it remains a fact that the signals provided by a woman's body shape do reflect her fertility. Women have to achieve a target ratio of fat to body mass both to undergo puberty and to ovulate. In addition, there is a relationship between WHR and fertility (the capacity to ovulate), reflecting the fact that fat deposition (in the hips, in particular) is regulated by oestrogen. Moderate levels of body fat also reflect general health. In a Finnish study, women with low WHR (i.e. tending towards 0.7, indicating a small waist and large hips) had lower blood pressure as well as lower cholesterol and triglyceride levels than women with high WHR (closer to 1.0). Similarly, in a sample of students in the slightly underweight to over-weight (but not obese) range, facial fatness correlated neg-atively with the frequency and severity of chest infections and higher blood pressure, and positively with a greater likelihood of having taken antibiotics. Waist–hip ratio also has consequences for aspects of infant development. In a study of 16,300 American mother–infant pairs, the moth-er's WHR was negatively correlated with her child's IQ (i.e. women with WHR tending towards 0.7 had higher-IQ children). This was especially true of teenage mothers. The authors argued that this was due to the availability of resources during fetal development.

The object of these kinds of cue is not just to make the girls attractive to the boys, but also to attract the boys to the girl so that she can pick and choose among them. We use all kinds of other tricks of the trade to do this. Besides the perfumes that we talked about in Chapter 2, there is the use of cosmetics to exaggerate the signals we have to offer. Cosmetics have an ancient history that dates back at least to 1400 BC, when three ladies of the court of the Egyptian pharaoh Thutmose III were buried with various jars of cosmetics, including a cleansing cream made from a mixture of oil and lime. Women of their social class enhanced their eyes by lining them with dark kohl (made from galena, a dark-grey lead ore, and still much in use today in Africa) or powdered green malachite. Eye shadows appear to mimic a natural darkening of the eyelids around the time of ovulation (probably the result of an enriched blood supply), and so seem to signal heightened fertility. Similarly, the use of rouge (probably mainly in lighter-skinned people) seems to mimic the way the cheeks are flushed during arousal.

The cues even extend to how you walk, the clothes you wear and how you interact during conversations. Karl Grammer, from the University of Vienna, whose lab has been responsible for generating most of the research on human courtship in the last three decades, filmed young women walking away from the camera and showed that they swayed their hips significantly more when they were ovulating than during the rest of the menstrual cycle. These women were in nightclubs and dance halls, and so, as we might say, on the make. In addition to filming them, Grammer and his team also measured the extent to which the women's dresses covered their bodies, and found that the amount of bare skin they were showing increased when they were ovulating.

Studies of human courtship carried out in singles bars have revealed a very distinct series of behaviours that women use when they are interested in a man they are talking to. These include leaning the upper body towards him, combined with a series of head tosses and hair flicks, and a distinctive pattern of laughter. These are often preceded by a momentary holding of the man's gaze, during which the pupils visibly dilate, followed by a coy look away. In his research on courtship, Grammer noted that women's behaviour seemed explicitly intended to control the development of the relationship, even though women seldom actually behaved negatively. If anything, women seemed to behave in a deliberately ambivalent way that gave the men few cues as to their real intentions. At the same time, they seemed to be probing for information to assess the man's suitability as a mate. In effect, they seemed to be trying to keep their options open as long as possible, reflecting their greater concern over ensuring the quality of any male they finally chose to mate with. Grammer suggested that this may explain why men frequently overestimate a woman's level of interest in them when they have only recently met.

However much we dismiss the importance of body shape, the fact is that we respond covertly to these cues in same-sex rivals as much as in members of the opposite sex. Doug Kenrick and his colleagues showed women subjects pictures of female models and measured their mood before and after doing so. After seeing the models, women's mood fell significantly: they were much less content with their own appearance. Similarly, when men were shown pictures of the same female models, they rated their wives as much less attractive and were less content about their relationship with them.

It seems that the folk wisdom about being tall, dark and handsome may not be so wide of the mark either, at least in the case of height. Boguslaw Pawłowski and I analysed data on the life histories of 4,419 Polish men from the city of Wrocław and showed that taller men were more likely both to be married than shorter men and to have more children, when education and place of birth were held constant. A similar relationship between stature and number of children was reported for the 1950 cohort of officer cadets at the prestigious US military academy at West Point (although in their case, the larger number of children was achieved by having more marriages). Daniel Nettle later looked for a similar effect in a very large cohort of UK males (the 1958 National Child Development Sample, which includes everyone born in England and Wales in one week in March 1958), but didn't find one; however, he did find a relationship between stature and the likelihood of being married. In a traditional society that did not have modern contraceptive practices, that would of course translate directly into more children.

In short, it seems that women actively select for taller men. Some evidence for that was provided by our Polish sample. It turned out that there was a secular trend in the magnitude of the relationship between male stature and their reproductive output. During the Second World War, Poland lost more men per head of population than any other European country. As a result, when the teenage cohort that had been through the war began to marry during the next decade, the women didn't have that many men to choose from. Consequently, evidence of a preference for tall men was virtually non-existent. But in each succeeding decade up to the 1980s, the effect became stronger and more explicit as women had more and more men to choose among.

When Nettle looked for the effects of stature in women in the 1958 UK birth cohort sample, he found a rather different pattern: a strikingly humped relationship in which shorter and taller women produced fewer children over their lifetimes than women of average height. The average height of women in this sample was five foot four inches, and the most fertile women were four foot eleven inches in height. This relationship held up even after controlling for socioeconomic class and the frequency of serious illness (since this is correlated with height in women). The reason for the low reproductive output at the extremes turned out to be that very short and very tall women were much less likely to be married or in a permanent relationship than women of average height. This suggests that humans are actually subject to conflicting selection pressures over height: women are selecting for taller men, but men are selecting for shortish women. Stabilising selection thus maintains our height as a species at a roughly constant level and prevents runaway natural selection driving it to either of the extremes.

There is, alas, an unexpected downside to the business of courtship. A recent study has shown that interacting with a member of the opposite sex impairs men's (but not women's) cognitive function. The test was a simple memory task: subjects were presented with a series of letters one at a time on a computer screen, and were asked to say whether or not each letter was the same as the last but one. Halfway through the task, they were asked to take a break in the room, where a male or a female confederate engaged them in conversation for several minutes. When they returned to the experiment and continued with the task, men who had interacted with a female confederate performed significantly worse than those who had interacted with a male confederate, and the decline was worse the prettier they rated the woman. The women subjects showed no such effect. Alas, it seems that men are easily distracted by a pretty girl and sent spiralling into a state of mental confusion.

Faces in the crowd

Men and women show striking consistency in their preference for masculine faces in men and feminine faces in women, and this preference crosses cultural and racial boundaries. Dave Perrett at St Andrews University has made a career out of studying facial attractiveness. He and his students have designed some very smart software which allows them to create composite faces that average the facial features of many individuals of the same sex, and then, having done so, allows the faces to be masculinised or feminised. The two sexes' preferred traits are quite distinct: women (and to some extent men) prefer men to have masculinised faces (a heavier jaw, wider cheekbones, more projecting lower face, larger eyes, prominent brow ridge, and darker skin tone, all traits influenced by high testosterone levels), whereas they prefer women to have feminised faces (smaller jaws, a narrow upper lip, less prominent brows, narrower cheeks, smaller eyes, lighter skin tones). You can subtly shift any face from more feminised to more masculinised, and back again, simply by adjusting these features in the appropriate direction.

The testosterone that drives these facial shifts along the masculine–feminine dimension is costly: it increases stress on the immune system and thus compromises immunocompetence, making males with higher testosterone levels less able to cope so well with attack by biological agents like viruses, bacteria and parasites. In that sense, masculinised features are hard-to-fake cues of good genes because they reflect the body's capacity to cope with the intense stress that testosterone production places on it.

By comparison to men's faces, women's faces are more neotenised (i.e. more baby-like). Babies have a very distinctive facial structure with a high forehead and a foreshortened face. As we develop, our faces gradually change shape as the jaws drop and start to jut outwards, becoming more prognathous. The ratio of the distance between the eyes and the top of the head relative to the distance between the eyes and chin is large in babies (basically, they are all brain) and then declines steadily with age. After puberty, the shape of a male's face changes faster, as the jaw line deepens and the lower face starts to protrude more to give the heavier, more angular male jaw.

In a study of women's facial features, Doug Jones measured various facial proportions, combined them into a single composite index and then correlated this index against age to produce a shape-for-age relationship. He then used this to predict the facial shape that an average woman of a given age would be expected to have. Comparing this predicted shape to a woman's actual shape gave a simple index of relative neoteny. When the faces were rated for attractiveness by subjects from five different cultural groups, he found that the ratings of female attractiveness increased as the difference between their predicted and actual ages increased: women whose predicted age was less than their actual age were considered more attractive. He then went on to do an analysis of the faces of magazine models. Compared against a standard shape-for-age graph, these women had faces that were the equivalent of a seven-year-old's. Their faces were exaggeratedly neotenised. In effect, facial neoteny appears to be a supernormal cue of youth, presumably the result of males choosing younger and younger-looking faces. Half hidden beneath this would seem to be a rather disturbing, if sad, explanation for paedophilia.

Because humans are so long-lived, it is rather hard to relate appearance when young to how well people do over their lifetime. This is particularly problematic for evolutionary studies, because what you would really like to know is how many children and grandchildren someone produces over the course of their reproductive life. Sometimes, quirks of circumstance provide just what you need. Sociologists Ulrich Mueller and Allan Mazur rated photographs of the officer cadets at West Point and related indices taken from their faces to their future careers in the military *and* to the size of their families, all of which were meticulously recorded in their alumnus records. First, they showed that men with more dominant (i.e. more masculinised) faces ended their military careers in higher ranks than those with less masculinised faces. Then they showed that those who did better generally produced more children *and* grandchildren. However, there was a cost to competitiveness. Those who made it to the highest rank of full general typically had fewer children: it was as though they overdid themselves and went too far, or the regime within which they were competing meant that in order to win the highest prize they actually had to sacrifice something.

However, by far the most surprising finding of all this face research has been the way women's preferences for male faces vary across the menstrual cycle. During the ovulatory (or follicular) phase, women prefer men with more exaggeratedly masculinised features, whereas at other times they prefer men with faces on the feminised side. They also prefer more symmetrical males during the follicular phase. Both highly masculinised traits and facial symmetry reflect gene quality – genes that are able to resist destabilising environmental effects during development. The explanation appears to reflect a 'cads versus dads' dimension (of which more in Chapter 7). In effect, when women are most likely to conceive, they shift their preference to men with cues of good genes, but at other times they prefer men with more feminised features because this reflects a more nurturing type. The strategy they seem to be playing is: conceive by the genetically best male you can find, and then get the most nurturing one to help bring it up.

One last curiosity discovered by David Perrett and his colleagues is that we tend to choose mates who resemble our opposite-sex parent, both in terms of facial shape and in terms of hair and eye colour. Similar results have also been reported from a Hungarian sample. Racial traits are an obvious case of parental resemblance since colour and hair form are such strong markers. Because of that, it provides a particularly conspicuous test case. And in fact, it turns out that children of mixed-sex marriages show a strong tendency to choose mates of the same race as their opposite-sex parent. In fact, so strong is this effect that children born of older parents even tend to prefer mates of about the age that their parents had been when they were young children. In other words, someone whose parents were, say, forty when he or she was born has a tendency to target older partners, irrespective of his or her own age, whereas someone who was born to parents who were in their twenties will be more likely to target someone younger. None of these effects are absolute, of course. Rather, they are statistical: but it happens significantly more often than you would expect if we really did choose our partners at random.

Faces do, of course, give multiple messages – cues of good genes in terms of testosteronised or feminised shapes, but also cues of relatedness and of personality. In a study of Canadians, Lisa DeBruine found that when people are asked to choose between photos that are opposite-sex transformations of their own face or averages of many unrelated individuals, they rated the self-similar faces as more attractive for a long-term relationship (the sharing, caring bit) but less attractive for a short-term fling (the good genes bit). This suggests that we can separate out the cues and use them in quite subtle ways to make choices for different purposes.

One of the more curious side effects of facial resemblance is what happens when babies are born. So great is the importance of bonding the father to the baby and persuading him that it really is his that the mother and the maternal grandparents will frequently comment on how the newborn baby resembles the father – at least, whenever the father is actually in the room. And this is a cross-cultural effect, having been demonstrated among North Americans as well as Mexicans. It's invariably facial features that come in for comment: how baby has his father's eyes, nose, mouth or chin; even foreheads and ears come in for their share of comment. Now the big issue here is that newborn babies don't actually look like anyone in particular. Family resemblances are features that only develop later, as babies grow up. In fact, facial features are probably designed by evolution to be neutral and not to look like anyone's in particular – just in case the father isn't the father. By the time the baby develops its full panoply of adult features and comes to resemble its biological parents, it has been living with its 'adoptive' dad for a decade or more and it's too late: 'dad' is well and truly bonded – which, of course, is why adoption works. The psychological pressure to make these comments is so strong you can observe it almost every time a new baby arrives. It's as though the mother and her family are desperate to convince the husband that the baby really is his.

There may be some real substance behind this. My former colleague Steve Platek asked men and women to rate a series of five photographs of a two-year-old that had been morphed with between 3 and 50 per cent of their own personal facial traits, and to say which one they found most attractive and which one they would lend money to or consider adopting. Both men and women showed a striking tendency to prefer faces that contained at least 25 per cent of their own traits. However, the effect was *much* stronger for males, approximately double that for the women. By comparison, faces with 3, 6 or 12.5 per cent of their traits were selected only at random. When asked afterwards, the subjects (all students at university) were unaware that any of the faces were morphs of their own. Their response seemed to be quite unconscious. In a subsequent brain imaging study, a small sample of men and women were shown photographs in which their own faces had been morphed on a fifty–fifty basis with either a two-year-old's or an adult's face. Men responded much more strongly to child morphs of their own faces (although women responded more strongly to child faces in general). In addition to the fusiform face area of the brain being active in both sexes (as one might expect on a face recognition task), men showed more activation in a number of frontal lobe regions (in particular) than women did. Not only did they seem more attuned to identifying their own traits in children's faces, they also seemed to be engaged in more active processing than women were. Men thought about the face more carefully, perhaps searching for cues to reassure them that the child really was theirs. I suppose the ultimate comment on how much faces mean to us is the extent to which some cultures go to hide women's faces. In many traditional cultures around the world, including Western Europe in the not so distant past, women cover their heads with shawls. More familiar in the present day, of course, is the fact that in some of the more extreme Islamic cultures women must be veiled from head to toe in public. One obvious reason for doing this is to reduce the risk of attracting men's attention by minimising the cues that are otherwise there to be read.

The sweet smell of symmetry

Symmetry of bilateral features (those parts of the body that have a mirror image, like the face, ears, arms or legs) is thought to be an important cue of gene quality. It seems to reflect how well the individual's genes can produce an exactly symmetrical body despite all the traumas and insults that life throws at us as we grow up. Since every cough and sneeze, every brief period of food shortage, every infection destabilises the developmental processes, genes that can cope with these bad circumstances and still produce the perfectly symmetrical you must be good. The differences are often tiny, almost imperceptible – a difference in the length of the ear lobes, or of the right and left index fingers. Yet somehow we seem to be able to pick them up. Over the past three decades, many observational and experimental studies have been done on symmetry (paradoxically, always referred to in the technical literature as *fluctuating asymmetry*), and on mate choice in relation to symmetry, in birds and other animals, and these have produced extremely robust effects. In humans, men (but not women, interestingly) who are more symmetrical are rated as more attractive by both sexes. On the other hand, breast symmetry seems to be correlated with fertility in women: women with more symmetrical breasts produce more children. Symmetrical men and women also have higher intelligence, and, which is even worse news, men with higher intelligence have more fertile sperm, whether this is measured by count, concentration, or motility (all of which influence a male's ability to sire offspring). Quite what causes this relationship is another matter.

Perfection, then, tends to come as everything or nothing, bad news though that may be for most of us. The bottom line, however, is that no matter what the evolutionary significance of symmetry may be, the reality is that it does seem to underpin human mate-choice decisions. And we are sensitive enough to it to be able to pick up extraordinarily subtle differences between individuals.

But it gets worse. Will Brown (then at Rutgers University in the USA) videoed men and women dancing in Jamaica and found a correlation between men's symmetry (measured as the mean difference in the sizes of left and right ankles and wrists, finger and ear lengths) and how well they danced (as rated by both sexes), with a much stronger effect in men. Presumably reflecting this, both sexes had a greater preference for symmetry in male dancers than in female dancers, and this effect was stronger for female raters than for male raters. In this case, women seem to be exerting a significantly greater selection pressure on men than men do on women, forcing them to push themselves to the limit so as to sort the men from the boys.

Although no one has ever looked at it, I have often wondered whether symmetry and physical gracefulness might play an important role in women's predilection for high heels. High heels have two important consequences for how women walk. One is that they force them to walk with knees slightly bent, because the heels tip them forward. That creates a tension right the way up through the body, and tension, like flushed cheeks, is indicative of arousal, and thus of one's sexual interest. It just makes you more alluring. But heels also have another effect: women naturally have wider hips than men, and heels make them sway more as they walk in order to maintain balance. That's difficult to do, especially when the heels are very thin, which perhaps suggests that heels might be a biological handicap. Walking elegantly and gracefully in high heels might be more difficult if you haven't got close to perfect symmetry in leg length and good co-ordination (and a good sense of rhythm?), thereby providing direct proof of the quality of your genes. It might also be a cue of age – something that only those with a supple, youthful body can do.

In an extensive series of studies carried out at the University of Albuquerque by Steve Gangestad and Randy Thornhill, more symmetrical men (those with low fluctuating asymmetry) were found to have had more sexual partners and to have engaged more often in extra-pair relationships. Neither the men's current resources and wealth nor their anticipated future salary predicted the frequency of extra-pair matings. Similarly, they found that more symmetrical men were more likely to be chosen as extra-pair mates by women who were in permanent relationships. As if it couldn't get worse, women are more likely to achieve orgasm with more symmetrical men. Symmetrical men even smell more attractive to women. And perhaps, given what we have already learned, it will be no surprise to learn that women found symmetrical men most appealing when they were ovulating.

*

So much for mate choice and our romantic relationships. But, as I mentioned in Chapter 1, we have other kinds of relationships with adults that, in their different ways, can be just as important for us and so shouldn't be overlooked. These are, of course, friendships and our relations with kin. In the next chapter, we'll explore these in more detail.

6

By Kith or by Kin

And here's a hand, my trusty fiere,
And gie's a hand o' thine;
And we'll tak' a right guid-willie waught,
For auld lang syne!*

'Auld Lang Syne'

We tend to make a big fuss of formal marriage arrangements. And we go into them expecting marriages to last forever. But in traditional hunter-gatherer societies, the kind in which we have spent most of our long evolutionary history, marriages are not always as formal or as long-lasting as this. Two people just decide to live together, and children arise naturally from that. When they have had enough, they part company and move on to find new partners. Among the Ache hunter-gatherers of eastern Paraguay, the average number of partners an individual has over the course of his or her lifetime is between ten and twelve. Even though many of these changes in partner result from the death of the previous partner, social arrangements are nonetheless quite fluid.

But just how different are romantic relationships from the other bonds that make up our complex social world? In this chapter, I want to compare them with two other specific kinds of relationships that we all have: family and friends.

* Fiere: fellow. Guid-willie waught: good-will drink. Auld lang syne: old times' sake.

Family and friends

Whether we like it or not, we are born into a network of kin. We all have parents and grandparents, and in all likelihood we also have siblings, aunts and uncles, cousins and second cousins. The layers of the kinship network sweep outwards from us by degrees of relatedness. Some will be kin by direct biological descent, others will be kin by marriage – in-laws or, as anthropologists tend to call them, affines. Although our affines are not related to us by blood – which has tended to result in them being ignored by evolutionary biologists – they share with our biological kin the fact that, by virtue of marriage, they have a genetic interest in our offspring. Biologically, that places them on the same footing as true kin (those with whom we share common ancestry).

There is, however, an important difference between modern post-industrial societies and traditional small-scale societies. In small-scale societies, whether they be hunter-gatherers of Paraguay, island communities of the Scottish Hebrides or rural mountain communities in North America, virtually everyone within one's social horizon will be kin in some degree. Your place in society is defined by your kinship relationships. They define how you should greet someone, whether and how you can joke with them, and – perhaps most important of all – whom you can marry. In modern post-industrial societies, however, our kinship web is interspersed with individuals who have no formal biological relationship to us. We call them friends, or, to use the old term, kith.* In one sense,

---

* Kith is a now obsolete word that derives from the Old English *cýþþ*, meaning kinsfolk or family, which after the Middle Ages came to be applied to friends in distinction from true kin with whom we shared a direct biological relationship.

our social world consists of two quite separate networks, a kinship network and a friendship network, and the two are intertwined. Our personal social networks consist of, on average, about half kin and half kith, but the balance in any given case depends mainly on how large your extended family happens to be. We prioritise kin. In our analyses of personal social networks from Belgium and the UK, my collaborators and I found that those who come from large extended families have fewer friends. It's as though the number of slots you have available in your social world is fixed. You fill these up first with family, and then, if you still have slots available, you add friends. There are at least two likely reasons for this.

One is the pressure of what biologists refer to as kin selection: an evolutionary pressure to favour kin because of the benefits that doing so typically provides in terms of propagating the genes that you share with them. When we share a common ancestry with someone, we share a proportion of our genes with them by direct inheritance from the common ancestor(s). In strictly monogamous mating systems, we share about half our complement of genes[*] with our siblings, and about a quarter with grandparents, while cousins share about an eighth. Biological theory tells us that these values weight the likelihood that any act of altruism towards a relative will yield a genetic benefit to

[*] Strictly speaking, it's that they have a probability of one half of sharing a particular gene, and that probability is constant across the full complement of genes. That more or less comes out to the same thing as sharing half your genes, but, for technical reasons to do with how genetics works, it doesn't always do so and the two ways of calculating relatedness can sometimes lead to different answers. However, as a rough approximation for present purposes, we can speak of siblings as sharing half their genes.

135

the altruist by promoting the replication of the genes the two of them hold in common. One of the consequences of this is that family are much more likely than unrelated friends to stand by us when we are in difficulties. Blood, as the common wisdom observes, *is* thicker than water, both literally and figuratively. The other reason why kin are prioritised is that family relationships seem to require much less effort to maintain than friendships. I want to say more about this second reason in the next section, so for the moment I will leave it at that. First, I need to say a little more about the structure of social networks.

If we rank all the people we know on a single continuum in terms of how close we feel to them, we typically find that they settle into a series of groups whose members share a similar level of emotional closeness to us, with relatively marked step-changes in closeness between the groups. In effect, they form a series of layers of friendship (in the Facebook sense). As we progress through these layers, from the innermost set of close friends to the outermost layers of passing acquaintances, two things become obvious. First, each successive layer is larger than the one inside it. Second, there is a close relationship between the level of emotional closeness we have to someone and the frequency with which we contact them, and both decrease as we pass through the successive layers. The layers begin with an inner circle of about five intimate friends and expand outwards to groups that number (including the layers within them) fifteen best friends and fifty good friends, to a limit at around 150 friends (now known as Dunbar's Number). Beyond this outer layer of 150 lie several further layers of increasing size that include mainly casual acquaintances and people we know by sight or name but may never have met and certainly don't have a real relationship with (including, for example, celebrities or TV personalities whom we would recognise in the street but who would almost certainly not know us).

The people in these layers don't necessarily have to be living beings. They can include dead relatives, famous people from the past, figures of religious significance such as Jesus Christ or the Prophet Muhammad (according to our particular religious affiliations), or the saints and gods of any religion, and even characters from our favourite TV soap. They may include our pet animals, perhaps even our favourite potted plants if we feel we have a relationship of some kind with these (which probably at least requires that we talk to them). It may well be that, if we include any of these virtual individuals at all, they are most likely to lie in the inner circle of intimates. But the point is that we can include whomever we like in our circle of friends.

These layers in our social networks exist for a very good reason: they represent groups of people that provide us with specific benefits. The inner layer of around five intimates represents our support network, the close set of individuals who provide us with emotional support, those to whom we go when we need advice (especially about crises in our personal lives) or financial or other help. These are the people who will come to our aid when we most need it. Several of these individuals will usually be close family members, but there will always be one or two intimates who are unrelated friends. Outside this sits an additional set of about ten individuals who, together with the inner core of five, act as a sympathy group, the group of people whose death tomorrow would leave us deeply upset. These fifteen individuals represent the set of people with whom we spend most of our social time, the people we see most of and do most with. They probably provide our main source of regular voluntary childcare when we need it, those who will come round most often when we need help moving or repairing something, the ones who will usually be our first port of call for barbecue or dinner invitations. Between them, these fifteen people account for something like 60 per cent of all the time we spend in social interactions. It is here, in these inner two layers of the network, that the real comparison with romantic partners will lie.

Beyond these first two layers lie the 130 or so individuals that we count as general kith and kin – distant friends, friends of friends, and distant relations, all of whom we would readily distinguish from the more anonymous relationships we have with those outside the magic circle of 150 that makes up Dunbar's Number. The people in these outer layers provide us with a wide array of services in terms of information about the outside world – knowledge about job opportunities, a pool of reserve friends should we fall out with any of those in our inner circles. To those who study social networks, these two outer layers are known as 'weak ties' because they are less close emotionally to us and are often indirect (friends of friends).

The importance of being networked

In recent years, there has been growing evidence that the size and cohesiveness of your social network can have a dramatic effect on your health and that of your children. Having friends, being married or being a member of a close-knit community like a religious group all reduce your likelihood of being ill, and speed your recovery when you do fall ill. In one study, the researchers found that when subjects were exposed to the common cold virus, those who had more friends were significantly less likely to go down with cold symptoms. In another study, wounds healed more quickly in patients with harmonious marital relationships than in those with more hostile relationships. Your social circle protects you in some way that we don't really understand. But one possible reason is likely the endorphins generated by interactions with close friends and family. Endorphins seem to tune the immune system.

In their book *Connected*, the Harvard sociologists Nick Christakis and James Fowler have shown in some considerable detail the important roles that our networks of friends play in our lives. We are more likely to become obese, to be happy or depressed, to give up smoking, to become divorced, even to die, in the future if other members of our close network do any of these now. They found significant effects in this respect out to the third degree of relationships within networks, that's to say friends of friends of friends. These wider circles of friendship can sometimes come to play a crucially important role in our romantic lives: roughly 70 per cent of us meet our future partners through a family member, a friend or a friend of a friend. In contrast, only 3 per cent of us meet a one-night stand through a family member, and only 37 per cent do so through a friend. In effect, the old-fashioned match-maker is still alive and well, just rather more of a group-based thing than it used to be. One reason for this is that our friends in effect stand surety for the prospective partner: they act as a guarantor of the individual's suitability as a life partner, thus reducing the risks of our making a badly informed choice. Surety is much more important for a long-term partner than for a one-night stand who is here today and gone tomorrow.

This is not the only circumstance in which we might rely on the advice of others in making our choices. There is considerable evidence from animals to suggest that females copy each other's mate preferences. It's as though they are using a collective voting system to find the most attractive mates. And there seems to be an equivalent effect in humans. If everyone else thinks that Mr Darcy is the cat's whiskers, then it is very likely that he must be the cat's whiskers. It's a cheap and cheerful heuristic that saves time and effort trying to find out for yourself. This is the so-called wedding-ring effect. Men who are already attached are seen as more attractive, especially for one-night stands (i.e. in the search for good genes). Casual observation also suggests that women discuss the merits and demerits of men among themselves far more often than men ever do in respect of women. In one recent study, forty young women were asked to rate photos of eight men for whether they fancied a date with them, and then to rate them again after watching videos of them at a speed-dating event. Since the speed dating was in Germany, and none of the women were German-speakers, the only cue they had was the apparent interest shown by the woman at the event. Their ratings of the men's desirability both as short-term sexual partners and as longer-term mates increased significantly after seeing them in videos where the partner in the speed-dating event was obviously interested in them, compared to when she seemed uninterested.

Whatever the importance of friends and networks, however, kinship stands out as something very different.

Aside from the fact that we prioritise kin above friends in our networks, we also prioritise family in a more explicit way. When we asked people to select a friend and a family member of each sex from each of their four main network layers (the layers of five, fifteen, fifty and 150) and then tell us how likely they would be to act altruistically towards them (say, by volunteering to donate a kidney to them), they were always more generous towards family than friends, irrespective of layer. And they were always more generous to those from the inner layers than those from the outer layers, irrespective of their relatedness.

Some years ago, we carried out a much more direct study of altruistic behaviour in which we asked people to perform a rather painful skiing exercise in which you sit with your back against a wall with hips and knees at right angles as though on a chair – but with no chair beneath you. The exercise is designed to strengthen the quadriceps muscles in the thighs and help you with those elegant hip-swinging slaloms down the piste. Initially, it's a very comfortable posture. But after about two minutes, it starts to become excruciatingly painful. After four or five minutes, most people collapse onto the floor. The only person in our studies who managed to hold the position for more than ten minutes (nearly twice as long as anyone else) turned out to be a 'resting' ballerina. We paid subjects a flat rate for every minute they could hold the position, and asked them to repeat the exercise up to six times for the benefit of a different pre-selected person on each occasion. These included themselves (with whom they obviously shared 100 per cent of their genes), a parent or sibling (with whom they shared 50 per cent), a grandparent, aunt or uncle (with whom they shared 25 per cent), a cousin (12.5 per cent), a same-sex best friend (0 per cent) and the charity Save the Children. The children's charity provided a baseline, something that everyone could warm to and on whose behalf everyone would presumably be motivated to work hard.

Biologists will not be surprised to find that the amount of pain people were willing to bear was directly related to their genetic relationship with the recipients. Others, however, will probably be surprised to learn that, almost without exception, the children's charity came off the worst by quite a long way. After that, the amount of pain subjects put up with increased as the genetic relationship to the recipient increased. Subjects invariably worked hardest for themselves. So much for our human aspirations to altruism, and the myth of all that nonsense about Sidney Carton in *The Tale of Two Cities* nobly going to the guillotine on behalf of someone else more deserving.

However, there were two other, rather more interesting findings that came out of this study (which, by the way, we repeated five times, in two radically different cultures on two continents). One was that women were generally much more generous all round than men: the slope of the relationship between the length of time they held the position and their relatedness to the recipient was not as steep for women as it was for men, and women seemed to discriminate less strongly between the members of their social worlds than men did. In addition, they were absolutely more generous to their friends than the men were, typically treating a best friend rather better than a cousin. This points, once again, to the fact that, in one sense, women are more intensely social than men, able to create deeper relationships even with unrelated individuals than men seem to do. The second finding was that the best predictor of kinship (and hence of willingness to bear pain on behalf of kin) was how much time subjects had spent with the recipient in the first decade of their life. Spending time with someone in those formative first ten years seems to create an emotional bond that is very hard to override, even decades later as an adult. This suggests that time invested in friendships determines their strength (on which subject, more below). Generally speaking, of course, we will usually have spent more time at this formative age with close relatives, simply by virtue of having been born into the families we have. And this perhaps also explains why we can bond as well as we do with adopted parents or adopted children, even though we are biologically unrelated to them. It's a reminder that, although evolution acts to maximise genetic outcomes, it does not – and cannot – work through genes at the proximate level of actual behaviour. What it does is select for biological proxies that have the effect of maximising our genetic fitness on the average and in the long run, even though on occasions that sometimes gives rise to anomalous results, such as being just as emotionally close to an adopted child or stepchild as to our own biological children – though I have to say that this doesn't actually happen as often as people like to imagine (step- and adopted children are, sadly, often the victims of discrimination).

We had originally wondered whether close kin and close friends were really one and the same thing, created by a strong emotional attachment over a lengthy period of time, and that it was just distant kin and distant friends (or friends of friends) who were different. In the end, all our various studies have persuaded us that kinship and friendship really are two very different kinds of things, seemingly underpinned by different emotions. There may be some similarities, in that both very intense friendships and very close kinship may well be influenced by the time you spend with the person, but kinship has an edge to it that friendship lacks. And this is especially true of distant kin. If a long-lost cousin turns up on your doorstep and asks for a bed, you will almost certainly welcome them in. If a friend of a friend did that, you might point them to the nearest YMCA.

While you may have no choice about the family you have, you can at least choose your friends. We and others have found that friendships tend to be characterised by similarity in likes and dislikes, known technically as homophily (love of similarity). It seems that the more things we have in common with someone, the closer the friendship will be. In our study, we identified five key traits that seemed to be particularly potent in creating friendships: having the same sense of humour, the same hobbies/interests, and the same moral values (and/or religion), having a similar level of education/intelligence, and having being born in (or, at least, grown up in) the same area. The more of these five traits you share with someone, the greater is your emotional closeness to them, and the more likely you are to help them out in time of need. These traits seem to be strictly additive: none of them is more important than any of the others, and it doesn't matter too much which combination of traits you share. Sharing, say, the same moral values, place of origin and hobbies is much the same as sharing the same sense of humour, educational background and place of origin.

Why friendships wane

Friendships are not stable and enduring things. They are fragile and subject to a steady erosion of quality if they are not serviced continuously. In this respect, they differ radically from kinship relationships, which seem to be remarkably stable through time and robust to almost any degree of neglect. Our extensive research into women's social networks in the UK and Belgium has revealed that the emotional closeness of a given friend in your network (measured on a simple 0–10 scale, from low to intense) declines in direct proportion to the rate of contact with the friend. This is not true of kinship relationships, for whom emotional closeness is unrelated to how often you actually see them.

We found much the same in an eighteen-month study of a group of sixth-formers from a Sheffield school during the year they were away at university. It didn't really matter whether they saw more or less of their extended family members over the course of the year, they remained just as close to them (in fact, they even seemed to like them better – absence makes the heart grow fonder and all that). But if they saw less of a friend over the year, they felt significantly less close to them at the end of it. In fact, the deterioration in relationship quality seems to happen rather quickly, within a matter of months of ceasing to see the person. That doesn't mean that you suddenly forget all about them, but rather that a steady and insidious decline in the relationship sets in. If you don't act to prevent that, it will, over the following year or so, result in that person dropping slowly down through the layers of your social network until eventually they drop out of your 150 and become one of the also-rans in the layers of acquaintances beyond. To maintain emotional closeness to a friend, the subjects in our study actually had to contact them more often during the second half of the study period, after they had moved away, than during the preceding months when they had all been at the same school and seen each other almost every day.

There are, however, marked differences in friendship style between the two sexes. A variety of studies on the social psychology of friendship over the past couple of decades suggest that, even though women are more social than men and typically have larger circles of friendship, they tend to have a smaller number of more intense relationships. Men tend to differentiate rather less between their relationships than women do. As a result, women's relationships are typically more fragile: when they fall out, they do so catastrophically. In contrast, when men fall out, they just go and have a beer together and everything is fine. My caricature of relationships that sums this up is what the world was like when you were about eight. At that age, girls typically have very intense relationships with particular friends: consequently, if Suzy doesn't invite you to her party, it's a major crisis, the end of the universe as we know it. But for boys, a relationship is standing on opposite sides of the road kicking a football backwards and forwards. I maintain that it doesn't really matter too much whether it is another little boy on the other side of the road or just a wall: as long as the ball comes back, that's a relationship.

We found a striking difference between the two sexes in the way that friendships were prevented from waning. In our long-term study, we had asked each person to complete a lengthy questionnaire at intervals in which they told us about each of the people they included as a friend, what they had done with them, when they were last in contact by phone or email, or face to face, and so on. When we looked at those friendships that had resisted the erosion effect over the eighteen months of our study, it turned out that what seemed to have been responsible for preventing relationship decline differed between the boys and the girls. For the boys, relationships that survived through time involved those individuals with whom they had actively done things together more often in the second half of the study than in the first half – things like going out socially together, playing sports, engaging in joint hobbies, going on trips together. But in the case of the girls, how often they did things like this made no difference at all. Instead, it was how much time they spent *talking* together (either face to face, on the phone or by texting) that made the difference.

This suggests that friendships are underpinned by rather different dynamics in the two sexes. For men, relationships are maintained by 'doing stuff' together, almost irrespective of what that 'stuff' actually is – though it is probably most likely to involve drinking, or attending or taking part in sports or hobbies. For women, on the other hand, what maintains the relationship is talking together. For women, the phone and social networking sites like Facebook are perfectly designed for their needs, but both are foreign territory to the blokes. I like to think that our results explain why women are happy to spend so much time on the phone but, as every woman knows, the average phone call lasts only 7.3 seconds for men (and especially so for teenage boys). After all, all they really need to say is: See you down at the pub at seven o'clock.

One reason why kinship relationships may be so robust is that, compared to friendships, they are usually more densely interconnected. In other words, with the possible exception of your most intimate friends, friends' relationships tend to be just with you, and not so much with each other – thus forming a starlike pattern with you at the centre. In contrast, kin tend to be more connected with each other, giving a denser pattern of interconnections. In addition, with kin we have the almost unique phenomenon of the 'kin keeper', the person (usually, but not always, a woman) who makes it their business to keep everyone up to speed on everyone else's latest circumstances. Thus, much less effort has to be invested by each of us to maintain our knowledge of the state of the kin network. All the gossip does the rounds through natural channels, so that everyone's impending marriages, births, deaths and doings are automatically widely known. With very rare exceptions, there is no equivalent 'friend keeper' to do the same job with the friendship network. We have to work at these individually.

This difference in network integrity may have important implications in terms of policing. In a kin network, there is a greater sense that one's behaviour is being watched, so you feel less inclined to behave in ways that others would regard as unacceptable. Information about your behaviour towards others is more likely to circulate through a densely interconnected network than one that is only sparsely interconnected. The importance of being watched over by *some*one has been nicely demonstrated by Melissa Bateson and her colleagues at Newcastle

University. They showed that when a photograph of a pair of eyes was placed above the honesty box beside a staff coffee machine, people more often put money in the box for their drinks than if it was a picture of a potted plant. And in a subsequent study, they showed this effect worked just as well in a large cafeteria with a notice exhorting users not to leave litter. The police in a large UK city tried this later, and reckoned that placing pairs of eyes on posters around the city centre reduced street crime by about 17 per cent. It seems that we are very sensitive to being watched.

That a greater sense of being watched might be important in networks became clear in a study carried out by Oliver Curry, another member of my research group. We were interested explicitly in the effect of network density on patterns of altruism, and asked around three hundred people to list their eight most important same-sex friends and to estimate the interconnections between them, as well as their own connections with each of them, measured in terms of how often each of them saw the others. Then we asked them to say how likely they would be to donate a kidney or lend £5,000 to each of the friends if asked. When we split the sample into those with the most and those with the least densely connected networks of friends – that is, those whose friends interacted most and least with each other, and not just with the subject – those who were embedded in densely interconnected networks were much more likely to say they would either donate a kidney or lend money to one of the others than were those subjects who had less well-connected networks. Reputation is important to us, and being found out as the one unwilling to help a good friend in need is likely to be damaging to one's reputation, just as much as not buying drinks when it's your round, or borrowing things and never returning them. What our friends might think gives us pause before we automatically reach for the 'no' button. In addition, this will in all likelihood be backed up by a greater sense of emotional closeness in well-connected networks, because these individuals interact more often and are more likely to feel that they are supported.

The costs of romance

In the previous section, I sketched out some of the differences between friendship and kinship. They are clearly very different kinds of things. But how do either of them differ from romantic relationships? Are romantic relationships more like intimate friendships? Or like close kin? In one sense, of course, the sexual element to romantic relationships places them in a different class from both. In the case of kinship, a mechanism known as the Westermarck effect, named after the Finnish anthropologist Edvard Westermarck, generally prevents matings between very close relatives (siblings, parents and offspring). For reasons that are still unclear more than a century after Westermarck published his findings, very close kin do not feel sexually attracted towards each other. It isn't 100 per cent effective (nothing is in biology), but it works well enough most of the time, which is all evolution worries about. The mechanism seems to depend on the individuals concerned having been in the same household when the younger of them was born, or at least was very young. The best-known example of this is provided by the child marriages that used to be traditional in Taiwan. The children were betrothed and formally married at a very young age, often while still toddlers, and they were then brought up together in one or other of the parents' homes. These marriages were notoriously infertile and often unconsummated, because the couple never felt any sexual attraction towards each other. Having grown up together, they simply treated each other as brother and sister. Some of the Israeli kibbutzim had the same problem when, inspired by liberal wishful thinking, they tried to raise children in communal nurseries rather than bother the parents with all that trouble. They had hoped that it would engender a deeper sense of allegiance in the second generation and so ensure that they married only within the community. Alas, it had exactly the opposite effect: all the children made concerted efforts to find spouses outside the community because none of them wanted to have sexual relationships with someone they had grown up with as a child.

The converse situation can, however, occur. Parents and offspring, or siblings, who were separated soon after birth often feel strongly sexually attracted towards each other when they eventually meet up again as adults. The experience is sometimes rewarding and enriching, and, needless to say, sometimes not. In part, this reflects a curious aspect of our natural mate-choice patterns. The likely explanation for this is that, in searching for suitable mates, we are looking for someone who has a broadly similar genetic make-up – not the same MHC complex, of course, as we prefer our mates to have a different MHC complex, but broadly similar in respect of the rest of their genes, the ones responsible for their general physical appearance. It is as though we are trying to preserve our own set of genes and prevent them being torn apart by the natural processes of biological reproduction.

A recent study of all Icelandic marriages since 1800 (Iceland is small enough to do a complete sample of everyone) revealed that the biologically most successful marriages – meaning those most likely to produce grandchildren – were between third cousins. More closely related partners (cousins, second cousins) or less closely related partners (fourth or fifth cousins and beyond) did significantly less well. In fact, there are good biological reasons for thinking that cousins of some kind will make the ideal mates. They share a significant proportion of our genes, but not so many that there will be a high risk of genetic disorders in the children due to inbreeding – the problem that geneticists refer to as inbreeding depression, meaning depression in fitness rather than any kind of psychological state. In effect, cousin marriages represent a balance in the trade-off between keeping together that perfect set of genes that you represent and avoiding inbreeding depression by marrying someone who is likely to share too many of the same debilitating recessive genes. Cousins (and more distant kin) can, and often do, feel sexually attracted towards each other. Cousin marriages are common in many parts of the world, and were common in Britain and France in the past, especially among better-off folk for whom it helped keep estates within the family.

Romantic relationships are not quite like family, but then they are not just like close friends either. They share similarities with both, yet are different. They are robust up to a point, in the way that kin relationships are. But equally, they are like close friendships in that they can fall apart – and when they fall apart, they invariably do so catastrophically. In one key respect, however, they are in a unique little category all their own. They are so much more intense than either kith or kin relationships that they can override both kinship and friendship.

Max Burton and I had around 540 people complete a questionnaire we drew up to allow us to examine the composition of the inner core of five most intimate friends in more detail than any of our previous studies had been able to do. As expected, the average number of close friends that people had was almost exactly five, split more or less equally between close family and friends. Fortuitously, Max had added, just as general background information really, a question about whether the person completing the questionnaire was single or in a romantic relationship. When we divided the sample into singles and the romantically attached, we were astonished to find that those in romantic relationships had one fewer intimate friend than the singles. They typically had only four intimate friends, including the romantic partner. We thought this might be because the intensity of romantic relationships takes away time and attention that would otherwise be directed at additional friends, and this suggestion receives some support from our analyses of how friendships decay.

However, we were struck by something else: if you think about it for a moment, it becomes obvious that the cost of falling in love is actually the loss of *two* close friends, not one, because by definition the romantic partner must have come in from outside this closed circle of privileged friends. You do not normally fall in love with someone who is already an intimate friend. Our data suggested that, on average, you lose one close friend and one member of the family from your inner circle when you acquire a romantic partner. That's a major cost. As we noted earlier, however, kinship is more resilient to stresses of this kind than friendships are, so while the relationship with the family member who gets demoted into the second tier will probably not be lost for ever, that with the friend who gets demoted will very likely start to deteriorate almost immediately. It looks as though, when we are forced to sacrifice two members of our inner circle, we try to limit the damage by imposing some of the cost onto relationships that will be more resistant to decay. If we dropped two friends, we would literally lose two intimates for ever, but by sacrificing a close relative we trade on the fact that they will be less upset by our ignoring them and will still be there in the future when we have more time to spare for them.

A small number of people in our sample declared that they also had a lover in addition to their official romantic partner (the legal husband or wife). (Curious what people will reveal in the anonymity of the Internet, isn't it?) Our interest was immediately piqued: would the costs of running two relationships be so great that they had to forgo *four* other relationships, ending up with an intimate circle of just three? To our surprise, the answer was a very clear 'no': those who declared an extra romantic relationship also had an average of four people in their intimate circle! Something wasn't making sense. Then Max noticed that in none of these cases had the official romantic partner been included in the magic inner circle. The reason why these individuals had a lover was that their primary partner had already been demoted from the circle of intimates into the second rank of plain friends. This seems to me strong evidence that we cannot easily run more than one genuinely romantic relationship at a time. They are too intense and too all-consuming of time, energy and emotion for this to be possible. When people claim that they can, it must be either that the second relationship is purely sexual with no emotional overtones or a case of neither relationship having the intensity of a full-blown romance.

When your choice is not your own

So powerful are the forces that create romantic relationships that they can even come between parents and offspring when the former disapprove of, or fall out with, the person that the latter has chosen to fall in love with. So many honour killings of daughters in Asian families, so many estrangements between parents and offspring, arise from these disputes. They are far from rare occurrences. They arise because parents have a vested genetic interest in the offspring that their children produce, so all through history and prehistory parents have tried to manipulate the marital arrangements of their children to promote those interests. Rightly or wrongly, parents take the view that their greater experience of life and the world gives them a better sense of what is, and what is not, a good match. I'll finish this chapter with three remarkable examples all drawn from real life.

In the late Middle Ages, the Portuguese nobility manipulated the life opportunities of their children to ensure the succession for the family estates. Previously, they had operated a system of partible inheritance whereby each son inherited an equal share of the family fortune. (Daughters gained a share too, but it was much smaller and mainly in the form of a dowry.) This had been possible because, as the Moors who had ruled the Iberian peninsula for the previous seven centuries were forced out, the Portuguese ducal families had been able to acquire their land for next to nothing. As a result, there was no shortage of land, and each generation of sons could easily make up a full-sized estate from the modest share they started with by taking over nearby Moorish estates. But once the Moors had all been expelled and there were no more of their estates to take over, the Portuguese nobility hit a problem: the size of estates became fixed, and dividing them equally between the children rapidly reduced their size and economic viability with each succeeding generation. A wealthy landed family who produced too many sons would soon end up with descendants owning tiny, uneconomic parcels of land. In short, within a couple of generations, they would have joined the peasantry. So they did the obvious thing: they switched their inheritance rule to primogeniture, and the oldest son inherited the entire estate. This, of course, created a surplus of younger sons with no prospects. Now, boys without prospects are troublesome the world over. And inevitably, that's exactly what happened: gangs of unruly younger sons rampaged their way around Portugal, behaving like bandits and becoming increasingly disruptive. It may be no coincidence that the era of Portuguese explorations to discover the New World and the sea route to India began at just this moment in history. What better way to be rid of troublesome youth than to encourage them to go and make their fortunes elsewhere at someone else's expense?

They had a similar problem with excess daughters as well, of course: if each estate is inherited by only one son, he needs only one wife from another landed family. What to do with the surplus girls? The solution they adopted was to place them in nunneries. These girls led a privileged life slightly apart from the other nuns with their more conventional vocations. They were known as Brides of Christ, entered the nunnery with dowries (it was a condition of acceptance) and could be removed at any time if an older sister died and the family suddenly needed a replacement for a strategic marriage.

The second example comes from the Krummhörn region of north-west Germany, whose good folk we met in Chapter 4 through Eckart Voland's seminal analyses of their marriage records. During the eighteenth and nineteenth centuries, the petty farmers of this region practised a form of ultimogeniture – inheritance of the family estate by the *youngest* son. The problem he faced was that he had to buy out his older brothers. Although they did not get more than a token share, it was often enough to force the inheriting son to sell off some of the land. Ultimogeniture helped reduce the frequency with which this happened by enabling the parents to retain control over the farm longer than they would have been able to do under primogeniture: the oldest son would have been pressuring them to hand over the farm as much as a decade earlier than the youngest son. But the canny farmers of the Krummhörn also had a joker card up their collective sleeves. In order to reduce the drain on the family inheritance, they manipulated the survival chances of their sons to reduce the number of sons they produced to the bare minimum – one son to inherit and a back-up in case he died, a tactic known among demographers as the 'heir and a spare' gambit.

Voland and I were able to show from the Krummhörn parish records that third- and fourth-born sons suffered mortality rates as high as 50 per cent in the first year of life – almost three times the death rate for first- and second-born sons. This was probably not the result of infanticide as such, but rather of neglect – just not looking after them as well, being less willing to spend money on things like visits to the doctor. Daughters suffered no such costs, since they could always be married off into other families. Their survival rates were unaffected by birth order. In effect, the Krummhörn farmers seemed to be deliberately manipulating the number of surviving sons they had to ensure that they didn't impose too much of a drain on the family farm when the inheriting son came to take it over.

Our final example concerns the Tibetans, and is based on ethnographic studies carried out on the inhabitants of the Ladakh valley (so-called Little Tibet) in the Himalayas by John Crook, who, many years before, was my original PhD supervisor. The Tibetans faced a similar problem: their high-altitude habitat is extremely poor in terms of agricultural productivity, and the population was, and largely still is, at maximum capacity. Rather than subdivide their already tiny family farms into completely uneconomic units, they had adopted a rare and nearly unique form of polyandry in which all the sons are married to one wife. The sons then co-operate (more or less) in running the farm, sharing the wife between them. On the whole, the arrangement is not really satisfactory for anyone: the women find coping with several husbands at best a trial, and, sooner or later, sexual rivalries inevitably unsettle the harmony of fraternal relationships. One indicator of these latter stresses is that, in contemporary times, younger brothers opt out of the farm and its polyandrous marriage altogether if they can, and find a wife of their own – but they can only do this if they have an independent income, usually from some form of government employment.

Tibetan culture has obviously been well aware of this threat to the stability of the family unit, because there is one more twist of cunning in the story: parents commonly placed the second son in a monastery where he became a celibate monk. Think about it: when the brothers marry, the oldest would typically be about twenty-four, the second nineteen, the third fourteen and the fourth maybe nine (with daughters occupying the spaces in between). Well, for at least a few years, the two youngest brothers would not be interested in sexual relations with their wife, and she wouldn't be much interested in them. But the second son would be a serious problem: sexual rivalry would soon loom large. By removing the second son altogether (and giving him the seemingly worthwhile job of keeping the gods on his side), the family buys time for itself. The oldest husband can get on with the business of siring children safe in the knowledge that his younger brothers won't be competing with him for the attentions of the wife. By the time the younger boys get to be interested, the oldest one is passing into a sexually quieter time of life when he will be less threatened by their rivalry. On the whole, the system works, though the real losers are, as always, the excess daughters who cannot marry: they end up becoming drudges in the family home.

The use of celibate religious institutions to absorb some of the more troublesome members of the family (the excess sons) is far from uncommon. Another student of mine, Dennis Deedy, examined the parish records of Irish farming families in County Limerick during the second half of the nineteenth century. He was able to show that boys who went to the local seminary to train as Catholic priests were most likely to have come from families with an unusually large number of sons. We cannot say how much pressure the families placed on their sons, or whether the sons were eyeing up the family situation and drawing their own conclusions, but the net effect was the same either way: the families were reducing pressure on the estate to ensure that the inheriting sons, and later the grandsons, did not become economically disadvantaged by ending up with farms too small to be viable. In this respect, the Catholic Irish contrast with the northern German population in the Krummhörn, who, being Lutheran, did not have the option of monasteries and celibate priesthoods as a place to offload their excess sons. They had to find an alternative option.

These examples show how imaginative humans can be in finding solutions to the problems of lineage survival. All three populations faced the same problem, and found essentially similar solutions, each culturally appropriate to their circumstances. They also show rather clearly the extent to which parents try to manipulate the life chances of their offspring to maximise the benefit to their genetic lineage, sometimes without their offspring even realising that they are doing this.

*

There has been a hidden thread running through this chapter: even though some relationships are more robust than others, relationships can and do fail, and when they do so they invariably fail catastrophically. In the next chapter, we will grapple with this issue head on. Why do some relationships fail?

7

A Cheat by Any Other Name

While we sit bousing at the nappy,*
An' getting fou an' unco happy
We think na on the lang Scots miles
The mosses, waters, slaps and styles,
That lie between us an' our hame,
Whare sits our sullen sulky dame,
Gath'rin' her brows like gath'rin' storm,
Nursin' her wrath to keep it warm.
                    'Tam O'Shanter'

Close relationships have an uncertain future, unless they involve family. For romantic pairings as much as friendships, there is an inevitable deterioration over time unless constantly reinforced by frequent interaction. That is, of course, always easier for romantic relationships, simply because the pair usually live together. However, unlike pet dogs (whose devotion seems to be unqualified), we seem to have an innate tendency to trade on our relationships, sometimes pushing them to their limit to gain advantage for ourselves. Pushed too far, even the most devoted relationship can reach the point of exasperation and collapse. And when that happens, it is often precipitous and calamitous. Or to quote that eternal cynic Samuel Johnson on the subject of matrimonial bliss: 'the excesses of hope must be expiated by pain; and expectations improperly indulged, must end in disappointment.'

The truth is, however, a bit more mixed. On the one

* Bousing: drinking. Nappy: hostelry.

hand, a satisfying relationship is crucial for both physical and psychological health. But on the other hand, time notoriously does not treat marriage well. The so-called honeymoon years typically last only between twelve and thirty-six months, after which the relationship settles down to a state of laissez-faire normality. Things shift from fun and excitement to the drudgery of everyday life, with an increasing tendency to take the partner for granted. Squabbles increase in frequency and intensity, satisfaction declines and the risk of relationship breakdown rises until, eventually, things fall completely apart.

Breaking up isn't hard to do

Close relationships don't necessarily last forever. A third, perhaps even as many as a half, of all contemporary marriages in Europe and America end in divorce. We asked 540 people to tell us whether they had recently broken up a close relationship. Of these, no fewer than 413 reported that they had fallen out with a total of 903 people in the past year. That implies that each of us has a serious falling out with around two of our close friends every year. That might be a bit on the high side, because it could well be that people were more willing to complete our questionnaire if they had recently fallen out with someone – and, indeed, it was the case that fallings out were more likely to have occurred within the previous month than during the rest of the year. But still, it does suggest that falling out with close friends is relatively common.

The most common person to have fallen out with, and this was perhaps no surprise, was a romantic partner. Nonetheless, these accounted for only around 20 per cent of all relationship breakdowns. A top-five friend was the next most common, accounting for around 15 per cent. The next most common was a parent, making up a further 10 per cent, followed by a brother/sister or a non-top-five friend, who each accounted for around 8 per cent. The rest of the break-ups involved a motley collection of relatives, in-laws, colleagues, housemates and leisure-club friends. Overall, however, relatives ranked high: 65 per cent of all relationship breakdowns were with a close relative – someone no more distant than a cousin.

Blood may be thicker than water, but you are forced into much more frequent contact with relatives and this inevitably places strain on the relationship. One of the reasons for this seems to be that we often take relatives, and especially close relatives, for granted, and so trample on their sensitivities more often than we realise. This probably goes back to the point that kin relationships are more robust than our rather fragile friendships. We have to work at friendships to keep them going, but we can let family relationships slide. It seems that sometimes we just overstep the mark once too often, and the result is a catastrophic collapse. With friendships, and especially less intimate friendships, things may be more likely to slide rather than collapse: they offend us, so we just don't bother to see them again, and the relationship naturally fades. But there is more at stake with a family relationship, partly because it inevitably has a long history and partly because our futures remain intertwined whether we like it or not.

Why did the relationships in our sample fall apart when they did? When we looked at the causes of the break-ups, it was clear that these centred around four particular kinds of event: insults (whether given in public or private), failure to be at an important event (such as birthdays), spreading lies or rumours, and remonstrations (scolding) over bad behaviour. Between them, these accounted for around two-thirds of all break-ups. Insults accounted for around a quarter of all break-ups, scolding accounted for around 20 per cent, and the other two for around 15 per cent each. The next most common was rivalry of some kind – being friendly to an enemy or competing directly for a sexual partner. Between them, these accounted for a further 10 per cent of break-ups. So threats to one's social status or self-esteem accounted for over three-quarters of all relationship breakdowns. In contrast, more practical causes such as unpaid debts only accounted for around 5 per cent of cases. In other words, most relationships seemed to collapse for reasons that were explicitly about the relationship itself.

Parents were much less likely to say that they had fallen out with their children than vice versa: 10 per cent of children reported that they had fallen out with a parent, but only 3 per cent of parents (and all of those were women) reported that they had fallen out with one of their offspring. In other words, how a relationship is perceived can be very asymmetric. You may go off in a huff and consider yourself as no longer having a relationship with me, but I may view this as a minor blip on the surface of an otherwise stable relationship. Women were nearly twice as likely to say that they had fallen out with a romantic partner than men were: rifts with a romantic partner accounted for 25 per cent of all women's break-ups, but only 15 per cent of men's. But men were twice as likely as women to have fallen out with a sibling, three times as likely to have fallen out with a colleague, and four times as likely to have fallen out with an 'extra' romantic partner (a covert lover).

It seems that the most dangerous point for a relationship is three years into it. Nearly half of all the breakdowns in our sample occurred between two and four years after a relationship started. Otherwise, it was long-term relationships (lifelong ones, or at least those that had lasted for ten years or more) that were most at risk. Interestingly, there was a sex difference here too. Men were more likely than women to have a relationship breakdown after three years, but women were more likely than men to have a breakdown in a lifelong relationship. You might think this is because men are generally less willing to work at their relationships, and more prone to spontaneous 'That's it, I've had enough!' reactions, but I couldn't possibly comment . . .

Sandra Murray and her colleagues followed nearly two hundred childless couples over the first three years of their married life. They found that the more an individual idealised their partner at the outset, the longer they continued to be satisfied with the relationship, and the more their partner idealised them, the better off they were. Those who thought especially highly of their partner still thought just as highly of them three years later, whereas those who idealised their partner less showed a steady decline in how they saw the partner, irrespective of how realistic their views actually were. In fact, the more unrealistic their view (i.e. the more idealistic they were about the partner), the more likely was relationship satisfaction to remain high. Perhaps contrary to folk wisdom, it seems that being unrealistic in evaluating your partner's good points doesn't necessarily lead to greater disillusionment when reality strikes. In fact, quite the reverse: the rosier the picture, the longer the rose petals will remain in place. So far from dampening enthusiasm, greater idealism actually predicts greater satisfaction. Or perhaps it's that the illusion just continues to survive better.

It's the pain that you feel

Poets and spurned lovers speak of the physical pain that rejection brings. Both the internal and external signs are much the same as in physical pain. Yet one involves physical injury, and the other is a purely psychological experience. In fact, it turns out that both kinds of pain are processed in the same part of the brain, an area known as the anterior cingulate cortex (or ACC for short), which lies just below the main layers of the cortex in the centre of the brain. People who have an unusually active ACC are particularly sensitive to pain, and individuals whose ACC has been damaged or destroyed are generally insensitive to pain. So much so, in fact, that surgeons often now deliberately destroy parts of the ACC in patients who have chronic, uncontrollable pain.

Now it turns out that exactly the same area of the brain lights up when you feel psychological pain. This was demonstrated in a rather neat, if unkind, experiment in which Nancy Eisenberger and her colleagues asked subjects to play a computerised ball-tossing game called Cyberball while lying in a brain scanner. They watched a computer screen on which two figures threw a ball to each other, with the subject lying in the scanner represented by a pair of hands at the bottom of the screen. Subjects were told that the figures represented other players in nearby scanners; by pressing a left or right key, they could choose which of these two to throw the ball back to when one of them threw it to the pair of hands. Initially, the two figures on the screen included the person in the scanner by throwing the ball to the hands, but after a few goes they just tossed the ball backwards and forwards between themselves. When the person in the scanner was excluded from the game, they showed greatly increased activity in the ACC. Just how much activity there was in the ACC was determined by how distressed they felt by the experience (as indicated afterwards in self-reports). Interestingly, the ACC is more active (and the perceived distress greater) when the individuals who shun you belong to the same in-group, whether that be the same race, the same social class, co-religionists or just people who resemble you (given that, as we saw in the previous chapter, we actively seek close relationships with the people who most resemble us in social traits and physical appearance).

It seems that the ACC's function is to alert us to situations where there is a discrepancy between how the world is and how we think it should be. In some cases, these will be logical discrepancies (things that just don't make sense, so we need to figure out what's wrong); in other cases, these will be injuries (so we need something to tell us to get away from the injurious situation); in yet other cases, it can be social distress caused by a mismatch between our expectations of how our relationships should be and how they actually are. What seems to have happened is that, during the course of mammalian evolution, the mechanism for alerting us to social situations has taken over the system for alerting us to physical pain because pain is a good way of making us pay attention and do something about our circumstances.

In addition to the ACC, the right prefrontal cortex also turns out to be active during social exclusion experiments. However, the frequency with which the prefrontal cortex is activated in exclusion contexts is inversely related to the amount of activation in the ACC *and* the amount of self-reported distress, which has led to the suggestion that it has an inhibitory or control function, damping down the pain signals from the ACC so as to avoid overreacting to circumstances. Activity in the right prefrontal cortex seems to be triggered by actively *thinking about* rather than merely *experiencing* events. Its function may thus be to manage the ACC and other emotional responses and prevent them getting out of hand. Interestingly in this respect, it is one of the few higher cortical areas that have direct neuronal connections to both the ACC and the amygdala (which handles responsiveness to emotional cues, and especially aggression). The involvement of the frontal cortex is perhaps significant, given our evidence that precisely this region is involved in mentalising abilities and the size of social network that people can maintain (as we saw in Chapter 3).

One way this pain effect might have become involved in social rejection is through the mechanism designed to ensure that infants don't get separated from the mother on whom they depend for both safety and, of course, milk. Infants of all mammal species, including our own, have distress vocalisations that are given both when there is physical pain (injury, wind, etc.) and when they are separated from the mother. Mammal mothers are highly attuned to infant crying and will respond immediately, especially to cries by their own infants. (Later, of course, the little blighters learn that this is a good strategy for getting their own way, and will shed crocodile – and sometimes real – tears in the sure and certain knowledge that those around them will relent. But that's another story.) Distress vocalisations are dependent on having an intact cingulate cortex. If this is removed surgically in adult monkeys, the animals will no longer give distress calls, whereas if it is stimulated electrically they will do so spontaneously.

Some further evidence that social and physical pain are one and the same thing comes from evidence that children who have recently been injured experience distress in response to separation from a caregiver more often and more intensely than children who have not. Similarly, adults who have chronic pain disorders also experience more anxiety about their personal relationships than healthy individuals and are more likely to have an anxious attachment style (one in which they constantly question the commitment of a partner). The reverse also holds. People who are well embedded in social support networks experience less physical pain when coping with cancer, surgery or childbirth. It even makes you better able to tolerate electric shocks.

There are some intriguing parallels between people with obsessive–compulsive disorder (OCD) and those experiencing relationship breakdown. They seem to exhibit much the same kind of obsessiveness, in particular the inability to get the offending person or circumstances out of your mind. For individuals with OCD, of course, the obsession is more usually with the circumstances they find themselves in – its dirtiness or the risk of catching dreadful diseases through germs that the rest of us leave behind on objects. It turns out that people with OCD often seem to have an unusually active ACC, which presumably causes them to see problems in life where in fact none exist. These individuals exhibit an exaggerated response in their ACC even to simple logical conflicts. There is also evidence that people who score highly on the neuroticism dimension of the personality scale also have an ACC that is unusually active in response to a simple, non-distressing discrepancy-detection task. Having a hyperactive ACC might thus predispose you to reacting badly to minor events in the daily life of a relationship, and so make you more prone to relationship breakdown.

Interestingly, compounds that are pharmacologically active in either physical pain (e.g. opiates) or psychological states like depression (antidepressants) turn out to be an effective treatment for the other condition: antidepressants alleviate physical pain and opiates alleviate mental distress and depression. Infant mammals that are given tiny doses of opiates when separated from their mothers vocalise much less and exhibit greatly reduced levels of distress. Not surprisingly in the light of this, the endorphin system seems to play an important role in the sense of loss and distress: when women were asked to recollect the death of a loved one or the break-up of a romantic relationship while having their brains scanned, they exhibited much less activity in the endorphin receptor sites, implying that they were *experiencing* more pain. Once again, this suggests that endorphins might somehow be involved in the way we connect with others.

The endorphin receptor gene *OPRM1* comes in two forms, A and G, that differ by a single nucleotide (the building blocks of the genetic code), but this single difference turns out to have quite a dramatic effect on pain tolerance. People who are homologous for the A form (i.e. who have two copies of it, one inherited from each parent) are less sensitive to pain than people who have at least one copy of the G form. Typically, G carriers require higher doses of morphine after surgery, for example. They also turn out to be more sensitive to rejection in the Cyberball game, and to show higher levels of activity in the ACC during the game. So we are back with the endorphin mechanism that we discussed in Chapter 2. When social engagement – and physical contact – trigger the release of endorphins, the receptor sites in the ACC and elsewhere are filled up and you feel contented and at peace. In the absence of these social processes, your mind state is more neutral. But if an emotionally negative event occurs – the death of a loved one, or a betrayal or abandonment – then endorphin release is blocked and the ACC receptor sites remain empty, so that you feel raw physical pain. In less important cases, the prefrontal cortex might inhibit the ACC response. In effect, the prefrontal cortex (the 'thinking' part of your brain) makes a value judgement, and says: Listen, it's not that big an issue, forget it. But in emotionally intense relationships, the prefrontal cortex mechanism is overridden because the relationship is too highly valued.

It turns out that women (at least up until menopause) have higher densities of endorphin receptors in the amygdala, the ACC and the prefrontal cortex, as well as in a number of other brain regions of less immediate interest. Women have about 25 per cent more receptor sites than men do in these areas, but this ratio is reversed in the amygdala (in particular) in postmenopausal women. This gives women of reproductive years a higher pain threshold than men (something that is rather useful during childbirth). However, men seem to have a higher uptake of endorphins in the amygdala than women, and this might suggest that women *feel* pain more deeply than men (whose amygdala activation is dampened by the endorphin uptake). If so, and given the fact that physical and social pain appear to be the same thing, this might suggest that women would feel rejection more deeply than men.

Cads v. dads

The mammalian reproductive strategy of internal gestation and lactation inevitably skews the two sexes' options. Once a female has been fertilised, there are no direct reproductive advantages to further matings with more males. She will maximise her chances of successfully rearing offspring by hunkering down and investing everything in the processes of gestation and lactation (and whatever socialisation of the young might be possible afterwards). Males are barred from this option, and, in most species of mammals, can do nothing to assist the female. As a result, the only option they have to improve their reproductive output is to mate with more females. It is for this reason that mammals mainly mate promiscuously (or at best have polygamous mating systems in which one male monopolises matings with a number of females). Very few mammal species are monogamous, in contrast to birds where the majority mate monogamously because the male can help the female incubate the eggs and feed the nestlings.

Of course, if males *can* contribute to the rearing process, then it *might* pay them to mate monogamously and stay with the female. But, once again, this is a trade-off. In effect, a male evaluates whether he would have more surviving offspring over a lifetime by staying with one female or moving on as soon as he has mated with a female in order to fertilise as many as possible. Some years ago, I used a mathematical model to show that, across the great apes, whether males pursue a social or a roving strategy depends on the length of the interval between successive births for a female, the number of females that typically travel together as a group, how dispersed these groups of females are and the male's capacity to search his home range (essentially, how far he can comfortably travel in a day relative to the size of the area he has to search). The longer the female reproductive cycle, the larger the female group, the fewer groups per unit area and the shorter the male's daily range, the more likely males are to be social and stay with a female group once they have found one. An example is the gorilla, whose males are obligately social (they always live in groups with females). At the other end of the spectrum, where females are solitary, males should opt for the roving strategy, just as they do in orang utans. Chimpanzee populations fall between the two extremes. Modern human hunter-gatherers, who live in large home ranges that contain just a few relatively large female groups, lie well out beyond the gorillas and should always be social with the males permanently residing in the female groups (and of course they do).

However, living with a female does not of itself necessarily commit a male to monogamy. Males still have the option of choosing between being a caring, sharing dad or a misogynous cad with a roving eye. I guess it is self-evident that human males vary across this spectrum. In a study of male sexual behaviour in French-speaking Canadians from Quebec, Daniel Pérusse found that about a third of the men were habitually promiscuous, even though 90 per cent of men were married. Only about two-thirds were monogamous (by which I mean that they were faithful to their current partner, whether or not they had many such partners in succession over a lifetime). Interestingly enough, in the analysis of vasopressin receptor genes in the Swedish twin study that I mentioned in Chapter 2, the proportion of males carrying at least one copy of the 'infidelity' version (and therefore, presumably, at risk of behaving badly) was 36 per cent, almost exactly the same as the proportion of promiscuous men as observed in the Quebec study.

On closer analysis of the Quebec data, I was able to show that these proportions were a perfect mirror of the pay-offs that the males following each strategy could expect to gain in terms of matings (and, hence, if there was no contraception, the number of offspring they might expect to sire). Promiscuous males gained two-thirds of all the matings in Quebec. When the frequencies of two options multiplied by their respective pay-offs are equal, you have an evolutionary equilibrium, because if males switch from one strategy to a more profitable one, they will each get a smaller share when the total pay-off is fixed. And that will cause the first strategy to be favoured once again. In other words, the ratio is self-correcting: if the population drifts away from the 33:67 ratio, the pay-offs of the two strategies will change in favour of the rarer strategy, and it will naturally readjust itself. It seems likely that the baseline ratio of the pay-offs is determined by the women's willingness to mate with men who are already attached to another woman and/or who are unwilling to commit. That looks suspiciously like women pursuing a 'good genes' mating strategy of the kind we discussed in Chapter 5.

So it seems that roughly a third of men might be predisposed to behave badly when the circumstances make this a viable strategy. This fraction is likely to be independent of the official mating system favoured or permitted by the local culture (i.e. whether monogamy is socially imposed, polygamy is permitted or it is just a promiscuous free-for-all). It is much more likely to be determined by the underlying balance in women's strategic interests between good genes and parental investment. Where paternal investment is minimal, the ratio will drift in favour of men behaving more promiscuously, whereas if paternal investment is crucial to successfully rearing offspring, the ratio will drift the other way in favour of more committed men. And for this reason, even if some men are genetically predisposed to behave promiscuously, this doesn't mean to say they will always do so: that will depend on how they evaluate the balance of the benefits between acting promiscuously and being faithful.

Women, of course, are not completely exempt from all this. One study of forty-eight American couples revealed that women whose MHC complex was more similar to their partners were more likely to reject their partner's sexual advances (when the frequency of sexual activity was taken into account), had fewer orgasms, had more extra-pair sexual partners and were more attracted to men other than their primary partners, particularly during the fertile phase of their cycles. In contrast, the study found no effect on the male partner's sexual interest or extra-pair sexual activity as a function of shared MHC profile.

What this brings us around to is the possibility that the major factor predisposing for relationship breakdown is the availability of alternative partners. No matter how predisposed an individual may be to wander, the benefits of doing so will be limited if alternative opportunities are few and far between. Scott South and his colleagues have demonstrated just such an effect of the local sex ratio on divorce rates in the USA. In one study of white American youth in their twenties, they found that the risk of divorce was a U-shaped function of the sex ratio in the local marriage market (the population of unmarried individuals of this age in the local area). In other words, as the marriage market's sex ratio became more male-biased (i.e. women were rare and men common), women were more likely to instigate divorce, with the converse being the case (men more likely to instigate divorce) when the sex ratio was female-biased (a surplus of women). Similarly, in a larger-scale study of the adult population as a whole, they showed that the sex ratio in the workplace was the best predictor of divorce rates, with divorce rates being lowest when the ratio was even. Again, the issue seems to be the availability of alternative partners who are more attractive than the current one. Interestingly, at the same time, several factors seemed to slow down the rate of divorce, including the number of children in the marriage, marrying at an older age and being a homeowner. Another good buffer against divorce is the husband having had more than sixteen years of education (and thus likely to be in a better-paid job). All these suggest that people weigh up the costs and benefits of remaining in a relationship rather than just acting precipitously. In this study, two factors actually increased the risk of divorce: the relationship being a second marriage or the couple having cohabited before marriage. However, my guess is that both of these are correlates of some other underlying factor (such as being a member of a church that discourages both divorce and prenuptial cohabitation) and are not themselves the cause of increased divorce risk.

Trust in what you see

If males of a particular kind are especially attractive to females, then they are likely to have more opportunity to engage in extra-pair sexual encounters. And that, in turn, means that they are more likely to be tempted to do so because of the cues they receive suggesting that they have a strong hand in the mating game. In Chapter 5, we noted that more symmetrical men were more attractive to women and had more sexual encounters, independently of any effect there might be from their social status or wealth. Indeed, more symmetrical men are more likely to be chosen as extra-pair mates by women who are in permanent relationships, suggesting that women are actively targeting them. Moreover, more symmetrical men also invest less in their long-term relationships than less symmetrical men do.

However, this may be something of a two-way process. In one of the Albuquerque studies, men were tested in pairs competing for a date with a real woman via a video link. Symmetrical men were significantly more aggressive in their pursuit of the woman. When asked to explain to the rival why the target woman was more likely to choose them for the date, symmetrical men were more likely to denigrate and belittle the rival. In essence, then, symmetry is inversely related to trustworthiness, at least in males. Stature might also play a role here: in the study of American military cadets that we saw in Chapter 5, taller cadets had more offspring at the end of their active reproductive lives not because their wives were more fertile and produced more, but because they married more often (having presumably been divorced in between).

Folk wisdom tells us that you can judge a man by his look, and more explicitly that you should never trust people who won't look you in the eye. This turns out to be true: we can judge people's trustworthiness by their looks. First, there is a specific area in the brain (the so-called fusiform face area on the top of the temporal lobe) which is populated by so-called grandmother neurons that are particularly sensitive to faces (among other things). Right from birth, we naturally attend to faces and gain a great deal of information from them about a person's intentions. We process facial cues very fast: typically, less than forty milliseconds' exposure to a face is necessary to read its emotional state correctly.

One of the important cues given on our faces is, of course, smiling. But smiles, like laughs, come in several different kinds. The two most important are relaxed involuntary smiles, with their characteristic 'crow's feet' wrinkles in the outer corners of the eyes, and 'polite' smiles that are under voluntary control (and so don't exhibit crow's feet). The involuntary kind are known as Duchenne smiles, and so voluntary smiles are known as non-Duchenne smiles. We respond very differently to these two kinds of smile because the first suggests someone who is open and generous, and the second someone who is faking it. One of my students, Marc Mehu, was able to show that in a face-to-face interaction between two people, we are more likely to give Duchenne smiles when engaged in a task that involves sharing than in a control task that doesn't. Furthermore, we are more generous towards someone who gives lots of Duchenne smiles; in contrast, the frequency of non-Duchenne (i.e. 'polite') smiles has no effect at all on our generosity. In a later experiment, he was also able to show that we perceive people who give more Duchenne smiles as being more generous (though not necessarily more attractive or more trustworthy).

Some faces are more trustworthy than others, even without smiles. In one experiment, subjects were asked to play a trust game in which they had to share a cash reward with another person whose photograph was shown on a computer screen. They were significantly more generous to faces they rated as more trustworthy than to those rated as less trustworthy. Angeliki Theodoridou (whose research I mentioned in Chapter 2) found that people's ratings of the trustworthiness and attractiveness of faces was highly correlated. Self-similarity may also be an important factor. Lisa DeBruine showed that people judged faces as more trustworthy if they were an opposite-sex transformation of their own face than if the faces were composites of large numbers of strangers (i.e. lots of other people's faces blended electronically into an average face). She suggested that this reflects the fact that we are predisposed to trust people who look most like us and that this probably reflects kin selection (the evolutionary mechanism that drives us to favour relatives).

Sometimes these effects can be quite complex. The psychologist Doug Kenrick and his colleagues found that women who judged their opposite-sex partners as being relatively dominant in their relationships towards others also judged them to be less committed to their romantic relationship, irrespective of the partner's physical attractiveness. In contrast, men thought that attractive non-dominant female partners were typically much less committed than less attractive non-dominant ones.

It seems likely that these judgements are made automatically by the brain. Aside from the fact that they are made very fast on first seeing a face, they seem to involve some of the circuitry involved in processing intentionality and theory of mind. Two areas of the brain seem to be especially important in this respect: the amygdala (an evolutionarily old area that processes emotional cues, especially fear) and the orbitofrontal cortex (an evolutionarily more recent area right at the front of the brain that is involved in both reward and social and intentionality judgements). Patients with lesions caused by strokes in the right orbitofrontal cortex have considerable difficulty making trustworthiness judgements. The same turns out to be true of patients who have lesions in the amygdala, although in this case it has to involve the amygdalae on both sides of the brain: if only one amygdala is damaged, you can still make correct trustworthiness judgements.

Tania Singer, Ray Dolan and his colleagues explored this in more detail in a brain-scanning study. Subjects were shown 120 faces and asked to specify age (high school versus university) or trustworthiness (yes or no) by pressing one of two buttons. Age provided the baseline against which changes due solely to social judgements (i.e. trustworthiness) could be assessed. Five brain areas showed up as being explicitly used in the trustworthiness task and not the age task, namely the amygdala, the orbitofrontal cortex, the insula, the fusiform gyrus and the superior temporal sulcus (STS). The fusiform area isn't that interesting: as we saw before, it is involved in face recognition. The STS, likewise, might or might not be so interesting, since it forms part of one of the visual pathways and seems to be involved mainly in decisions about motion (and thus might simply reflect the brain's attempts to reconstruct the facial expressions illustrated in the photos, since in real life these are ultimately created by motion in different parts of the face). This may be related to the fact that smiling is being used as a cue for judging trustworthiness: the study found that happy faces were more often judged to be trustworthy, while sad and – in particular – angry faces were both judged to be trustworthy much *less* often. That said, however, the STS lies between the temporoparietal junction and the temporal pole, both of which, as we saw in Chapter 3, are involved in judging theory of mind and intentionality. The most important finding, however, is the strong effect they found in the amygdala, the insula and the orbitofrontal cortex (though in the latter case, the effect was specific to trustworthy judgements – there was no relationship with untrustworthy judgements), while the insula showed the reverse effect (it responded only to untrustworthy faces). In everyday life, the insula seems to be principally involved in processing and registering the body's autonomic responses, in particular disgust and other non-flight/fight emotions – in other words, the body's gut reactions to events 'out there'.

Mad, bad and indescribable

One of our less creditable gut responses in the romantic domain seems to be the old green-eyed monster, jealousy. Jealousy probably exists to protect relationships. In a sense, it seems like a good first try if a relationship looks as though it might be at risk: behaving in a possessive way might just be enough to bring a straying partner back into line. Hamadryas baboon males are so jealous that should one of their females so much as allow another male to get between them, even completely by accident, the male will launch a savage attack on her, with vicious bites to the nape of the neck. The females learn to follow their males come what may and to be very sensitive to straying, even accidentally. And it seems that we may be no better. We habitually use the threat of violence in almost any walk of life you can imagine – the bully at work, the thrusting young man determined to get his own way right or wrong (and, believe me, I've seen a few of them close up in my time). Needless to say, violence and the threat of violence rears its ugly head in the context of relationships too, both in romantic relationships and in close friendships.

Despite the fact that we abhor such behaviour and like to think of ourselves as civilised, aggression and violence are never far beneath the surface, especially in the context of romantic relationships. Martin Daly and Margo Wilson analysed the pattern of spousal murders and found that in 80 per cent of cases where husbands murdered wives, it was because of known or suspected infidelity or the threat of leaving. Outright murder simply represents the extreme case of the more general phenomenon of coercion. Coercion often seems to be a male's first response to both the threat of infidelity and the threat of being abandoned, and especially so in more patriarchal societies where men have absolute control over women. In the not so distant past, a Walbiri Australian Aboriginal man could beat, or even spear to death, his wife for the slightest complaint or neglect of duty, never mind the threat of leaving him, with complete impunity. Neither blood money nor even a public rebuke was due if a man killed his wife. Such seemingly casual acquiescence in spousal murder is by no means confined to traditional societies like the Walbiri. Under Anglo-Saxon common law as currently practised in Britain and the USA, murdering an adulterous wife is sufficiently excusable to warrant a lesser penalty, and much the same applies in Roman law as practised in France (the iconic *crime passionelle*).

In many societies, men use the claustration of women to reduce the risk of infidelity. This can range from confining women in harems to cultural restrictions on women's ability to travel. In traditional Sardinia, for example, women were effectively confined to their village for fear that they would catch malaria or the evil eye in the uncleared land away from it, even though men might travel widely to distant pastures with their flocks. Avoiding malaria is undoubtedly beneficial, especially for pregnant women, but one cannot help noticing that such cultural injunctions also had the benefit of ensuring that the women weren't given too many opportunities of meeting unrelated men. In traditional China, upper-class girls had their feet bound in ways that resulted in complete deformity. While the superficial appearance was delicately small feet (large feet in women are a turn-off for men of all cultures), one consequence was that they could not walk – hobble might be a better word – more than a few hundred yards. In short, it prevented them straying.* In some of the stricter contemporary Islamic societies, women are banned from leaving the home unless chaperoned by a male relative. Even then, they must be swaddled in outer garments – variously, the burqa and niqāb – that conceal their faces as well as their bodies from the gaze of men. Such restrictions are by no means the prerogative of Islam, of course. The Nambudiri Brahmins are a small, rich, exclusive caste of priestly landowners on the Malabar coast of southern India. Even as late as the 1960s, their womenfolk were only allowed out in public if their bodies were completely enveloped in clothing and they carried an umbrella to hide their faces. A bride could not even attend her own wedding: instead, she was replaced by a low-caste Nayar girl while she watched the proceedings from seclusion. Nor is the risk of one's women straying necessarily an exclusive concern of more traditional societies. In a large sample of 12,300 Canadian women, Margo Wilson and Martin Daly found that around 6 per cent reported that their husbands actively tried to prevent them seeing other men or tried to limit their social life in various ways.

These restrictions on women's ability to meet other men are, as much as anything else, a reflection of men's distrust of both their fellow men and their womenfolk. Beneath this, of course, hovers the hoary ghost of paternity certainty. If women have too much freedom and too much

---

* This practice was confined to the upper classes, who had more to lose in terms of family 'face' and wealth if their women strayed. Conversely, women played such an important economic role on the farm that the peasant classes could not afford to cripple their women in this way.

sexual licence, men risk investing in other men's children as a result of being cuckolded.

Much, however, depends on how effectively men can control women's sexuality or, alternatively, on how confident they can be that their women are not cheating with other men (in other words, on their perception of women's sexual predilections). In some societies, men are simply unable to prevent women having liaisons, and in these cases men invariably give up any pretence of worrying about paternity certainty and invest their time, energy and wealth in their sisters' children (whose relationship to them they can at least guarantee). These societies are almost all matrilineal rather than patrilineal. Well-known examples include many Polynesian societies – as Fletcher Christian and the other *Bounty* mutineers discovered to their undisguised joy.

However, when men can control women's sexual behaviour, the risk of adultery looms large in their psyche. Among the Afar nomads of the Horn of Africa, any man who stares too long at a woman is likely to be instantly killed by the husband or, if she is an unmarried girl, by her father or brother. Such societies often have a culture of family honour that views both premarital sex (at least by the girls) and adultery as so offensive that the only proper outcome is death. On a cross-cultural basis, women have fewest extramarital affairs in agricultural societies where male lineages often own the land, social stratification is highest and males are most overtly aggressive; conversely, they are most often involved in extramarital affairs in societies that practise shifting horticulture, where land is rarely owned, polygyny rates are highest and men contribute least to childcare.

To try to understand why a predisposition to behave violently might persist in a population, we analysed the family sagas of Iceland. Most of these were composed during the early thirteenth century as histories of individual families. Broadly speaking, they are thought to be reasonably reliable accounts of the events that befell particular Icelandic families or communities during the eleventh and twelfth centuries. Among the Vikings as a whole, some men acquired the reputation of being what amounts to sociopaths. Known as berserkers (from which we get the English verb 'to go berserk'), they were invariably physically powerful, renowned fighters and fearless in battle, with a tendency to strike first and ask questions afterwards. (Oh, and not infrequently they helped this along by taking dodgy potions brewed from various plants that have hallucinogenic properties.) Mostly the population lived in terror of them, but a berserker was frightfully useful to have along with you when it came to a spot of raiding on the shores of merrie England. Indeed, many of the Scandinavian kings liked to have them as their personal bodyguard (the dozen or so men who surrounded the king during battle). The problem was that they were invariably trouble back at home, as often as not using their reputations and physical strength to relieve other members of the community of their property, instigating vendettas that rumbled on for generations in the process. Our analyses of the family pedigrees has revealed that having a recognised berserker in the family dramatically reduced the frequency with which other males in the family were killed. Partly because of this, berserkers contributed more copies of their genes to future generations: not only did they reproduce more successfully than the average man themselves, but so did their male relatives. It paid to have a tame thug around. And it paid to be that thug, as well. In short, for better or for worse and whether we like it or not, there has been evolutionary selection for male violence and we are stuck with it. Our problem, as much as it was the Vikings', is how to control it.

Interestingly, it seems that we can judge a habitually violent person by his face, at least in the case of men. In one study, subjects were shown photos of eighty-seven registered sex offenders for just two seconds each. Since not all sex offenders are violent, the men naturally varied widely in terms of their histories of violence and that propensity to violence was written on their faces. On the basis of this brief glimpse, subjects made surprisingly accurate judgements of their propensities to violence. The best predictors seemed to be the degree of facial masculinity, the heaviness of the brow ridges and facial cues of physical strength. Notice that all the males were convicted sex offenders, so they weren't just picking out people who had been in prison or had a criminal history.

Research on psychopaths suggests that they have some other distinctive behavioural and neurological features. Clinically speaking, they are typically narcissistic, impulsive and manipulative, have a reduced responsiveness to emotional (e.g. frightening or fearful) cues, a complete lack of any sense of remorse and an inability to empathise with others. At the neurological level, psychopathology seems to be associated with a smaller than normal amygdala and reduced amygdala activity. In addition, the neural connection between the amygdala and the prefrontal cortex is less strong, implying a reduced ability to inhibit emotional responses as well as a reduced capacity to notice distress in others. This connection may also be important in allowing us to simulate others' distress in terms of our own, and if we can't do that we are likely to be less sensitive to others' suffering. Indeed, many of these same symptoms are present in individuals whose prefrontal cortex (and in particular, the orbitofrontal and ventromedial cortices) has been damaged by strokes or accidents. Similar effects can even be detected in normal individuals who lie on the high side of the psychopathic personality scale (i.e. individuals who show no *clinical* symptoms of psychopathology but express some of the underlying predispositions). The neural connection from the frontal lobe to the amygdala seems to be especially important in enabling us to moderate the natural tendency for the red mist to rise as soon as we are challenged. Without that controlling impulse, it's the hit-first-and-think-afterwards syndrome of the Viking berserker.

Incidentally, there is evidence that, even in healthy middle-aged women, high levels of testosterone have an adverse effect on how well the amygdala–frontal lobe connection works. It turns out that, related to this, there is a sex difference in the extent to which the left and right amygdalae connect with the orbitofrontal cortex on the *opposite* side of the brain (this is much reduced in men), although there is no sex difference in the connection with the orbitofrontal cortex on the *same* side. Because of this, highly testosteronised males (and that probably includes psychopaths) may simply be less capable of modulating their emotional reactions to threatening situations. The amygdala plays an important role in the conscious evaluation of biologically meaningful stimuli, and especially threats, so this might explain why, generally speaking, men are less responsive to (i.e. panic less about) threatening situations and are less often spooked by unexpected noises.

This sex difference extends to the way aggression is expressed in the two sexes. As the evolutionary psychologist Anne Campbell points out in her book *A Mind of Her Own*, men tend to respond to being threatened or challenged with physical aggression, whereas women tend to respond with verbal or psychological aggression. This has consequences down the line. Because their responses tend to be physical, of limited duration and less psychologically damaging, men seem able to resolve conflicts between friends more easily than women. Because women's conflicts tend to be expressed verbally and more often involve insults, the feeling of psychological hurt is greater and this is less easy to forgive and forget. As we saw in the first section, personal insults account for a quarter of all relationship breakdowns.

The two sexes also differ in what arouses jealousy. The balance of the evidence from a variety of studies suggests that men are typically more likely than women to become jealous if their partner has a fling with someone else; perhaps as a result, they are less forgiving than women. Women, on the other hand, are typically more concerned if their partner becomes emotionally involved with another woman – even if they don't have sex. In effect, women seem to become seriously worried only if there are resource implications because their partners will have to share their income and wealth between two families. It's the polygyny threshold problem again (see Chapter 4).

However, even within the same sex, we are not all equally susceptible to the green mist. Will Brown and

Chris Moore measured the bodily symmetry of a group of men and women and asked them to rate the amount of jealousy they typically felt in a relationship. Since symmetry is supposed to be a good index of gene quality (only the best genes can manage to produce perfectly symmetrical bodies in the face of the destabilising chaos of the natural environment), more symmetrical people should be more sought after and so more confident of their abilities both to retain a mate's loyalty and to attract substitutes if a mate does abandon them. As a result, they should be much less susceptible to jealousy. Conversely, since more asymmetrical people are less attractive, they should feel more threatened by rivals. The data showed that less symmetrical people were indeed more riven by jealousy than symmetrical ones, and this was true of both sexes. In contrast, non-romantic jealousy (i.e. envy of others' material success, such as a promotion at work) did not correlate with symmetry in either sex, and we would not really expect it to: symmetry is unlikely to have any effect on success at work. In other words, romantic jealousy and envy of others' economic or physical success are two quite different things.

The neuroscientist Tania Singer and her colleagues discovered a striking sex difference in how the two sexes' brains respond when observing someone being punished. Both sexes exhibited an empathy-like response in the main pain centres of the pain (the insula and the ACC). However, if the person being punished had previously played unfairly, men exhibited a somewhat reduced response in the pain centres and a strong response in the reward centres in both the left orbitofrontal cortex and an area known as the nucleus accumbens. An earlier study by Dominique de Quervain and her colleagues had shown that this area was activated when a subject was able to punish someone who had violated a social norm in a game, suggesting that its activation reflected satisfaction at a freerider being punished for their misdemeanour. Women did not show this effect at all, but rather responded with activity in the pain centres in both cases. Women invariably feel empathy with a victim of punishment, but men are inclined to wallow in a sense of *Schadenfreude* when they think the victim deserves it. Seemingly, berserkers and psychopaths are born and not made by circumstance.

*

So far, we have focused on the processes that underpin regular everyday relationships. But there is one class of relationship that bears many similarities to romantic relationships yet doesn't always involve the sexual overtones we associate with the latter. These relationships typically involve attraction to charismatic leaders, who can just as easily be religious figures as they can real humans. What makes such relationships fascinating is that they have many of the hallmarks of deeply committed romantic attachments, and can easily spill over into overtly sexual relationships as a result. Because they represent something of the extreme end of the spectrum, they provide us with fascinating insights into the dynamics of more conventional relationships.

8

Sleeping with the Devil

There's some are fou o' love divine;
There's some are fou o' brandy;
An' monie jobs that day begin,
May end in houghmagandie.[*]

'The Holy Fair'

In 1652, the Italian sculptor Gian Lorenzo Bernini put the finishing touches to his latest commission in the Cornaro chapel in the rather undistinguished church of Santa Maria della Vittoria on the Via XX Settembre in central Rome. His *Ecstasy of St Teresa of Avila*, an elaborate grouping of the Spanish mystic swooning before an angel after a vision of the face of God, was to become one of the high masterpieces of the Italian Renaissance. To be honest, Bernini had been in a bit of a pique when he made it. He had been the favoured artist of the Barberini family, and in particular of Cardinal Matteo Barberini when this august personage was elevated to the papal throne in 1623 as Pope Urban VIII. When Urban died in 1644, his successor, Innocent X, was of an altogether different cut. He was less enthusiastic about Bernini's skills and much preferred those of the sculptor Francesco Borromini. The papal commissions dried up and Bernini simmered in his studio. In due course, the Venetian cardinal Federico Cornaro began to ponder his own mortality and think about preparing a suitable tomb for himself. So he commissioned Bernini to sculpt an uplifting piece of artwork

[*] Fou: full. Houghmagandie: fornication.

for the church where he planned to be buried. Some say that Bernini's intensely erotic setting of St Teresa's vision of God was a deliberate thumbing of his nose at a pope who had eschewed his talents in favour of his arch rival.

Be that as it may, Bernini's *Ecstasy* is not only exquisite as a sculpture, it is also indicative of something far more interesting – namely, the fact that religious ecstasy bears more than a passing resemblance to the more secular forms of being in love. It has all the same hallmarks – the obsessively focused attention, the dreamy 'faraway' appearance, the internal anguish, the disinterest in worldly things like food and even sleep. At least within the Christian tradition, there has long been a tradition of mystical love for saints, the Virgin Mary, even God himself. Such relationships are not sexually consummated, of course, but they have that same feel of unrequited love.

In love with God

The mystical tradition in Christianity has a very long history. Barely a century after the crucifixion of Jesus, Montanus developed an ecstatic form of Christian mysticism in the deserts of North Africa. According to Montanus, ecstatic states gave humans direct access to God. Humans were, he opined, no more than lyres that God strummed. Though Montanus was regarded with inevitable suspicion by the mainstream hierarchy in the cities, the centuries that followed witnessed a veritable flood of individuals who took to the desert and its mystical experiences, sparked by his example. So important did many of these ascetics become that they eventually acquired the title of the Desert Fathers, among whom St Anthony was but the first among many equals. Here, in the shimmering light of the desert, they meditated and developed mystical practices that would probably not have come so naturally in the dank, rainy climate of Europe's crowded towns.

It was a movement destined to grow and prosper throughout early Christendom. Among the names associated with this mystical tradition in later centuries are the thirteenth-century Dominican friar Meister Eckhart, the iconic St Francis of Assisi, the twelfth-century Hildegard of Bingen (she of *I am but a feather on the breath of God* fame, and composer of much fine music), the fifteenth-century Englishwoman Margery Kempe, and, of course, St Teresa of Avila herself. In our own times, the tradition continues with the Capuchin friar Padre Pio and the late nineteenth-century Italian St Gemma Galgani. Read any of their writings and you have the clear sense of their being deeply in love with Jesus Christ himself. Here is the nineteenth-century St Thérèse of Lisieux in her *Story of a Soul*: 'How lovely it was, that first kiss of Jesus in my heart – it was truly a kiss of love. I knew that I was loved, and said, "I love You, and I give myself to You for ever."' And then, later:

> In a transport of ecstatic joy I cried: 'Jesus, my love, I have at last found my vocation: *it is love.*' . . . I have used the term 'ecstatic joy' but this is not quite correct, for it is above all peace which is now my lot; the calm security of the sailor in sight of the beacon guiding him to port. Ah! Love, my radiant beacon light, I know the way to reach You now, and I have found the hidden secret of making all Your flames my own!

It is impossible to read *The Story of a Soul* without being struck by the overwhelming sense of passionate, unrequited love, of an extraordinary desperation to be with and please that one special person, to do things for him, to bear endless sufferings, small and large, as gifts for him.

It is not hard to see the ecstatic aspects of religion as spilling over from conventional everyday falling in love. The difference is simply that we are falling in love with someone who doesn't actually exist – or who, in the case of someone like the Virgin Mary or many of the saints or even Jesus himself, did exist but are long since dead. Somehow the same buttons are being pressed, perhaps because when we fall in love with a real person we are actually falling in love with a fiction of our own minds.

Alas, some have taken all this far *too* literally. Tanchelm of Antwerp was a hugely successful (indeed, much idolised) itinerant preacher who attracted enormous crowds in the Low Countries during the late eleventh and early twelfth centuries. So besotted was he both with his own fame and with the Virgin Mary, he eventually held a magnificent ceremony in a field attended by thousands of his followers to celebrate his betrothal to the good lady. (Unfortunately, it seems that she was busy elsewhere that day, so a sacred statue was conveniently found to stand in for her.) He was the medieval equivalent of today's televangelists, able to whip up enormous crowds into a fervour of excitement.

This sense of being in love with God is not unique to Christianity: it occurs in the other Abrahamic religions. Nothing is more exquisite than the love poems that provide the core to the qawwali tradition in Sufi Islam. Sufism represents the mystical dimension of Islam, and, like most mystical sects in the Abrahamic religions (the

Kabbalah in Judaism, the Gnostics in Christianity), was regarded with suspicion leading to outright persecution by mainstream sects. The qawwali tradition of Pakistan and Iran, exemplified by its greatest recent exponent in the form of the late, great Nusrat Fateh Ali Khan, is a form of singing that uses driving rhythms and ancient love poems to bring the singer and listeners to a state of ecstatic joy. Many of these poems are traditional ghazals, an ancient poetic form dating back to the pre-Islamic Arabic of the sixth century that expresses both the pain of loss and separation and the beauty of love. What gives the ghazals their power is the ambiguity inherent in their words, for the poems can be read as carnal love poems or expressions of an ethereal and pure love for God.

> *Kali kali zulphon ke phande nah dalo*
> Oh you with such beautiful, long black hair,
> don't ensnare me in your bewitching net,

pleads the singer, adding:

> The glance that pierces
> tranquillises the heart;
> spread the shadows of your tresses
> to make the darkness pleasing.

And again:

> *Meri ankhon ko bakhshe hain aansoo*
> You have brought tears to my eyes . . .
> I asked for love but received only sorrow.

Or, in the slightly coy words of the Old Testament 'Song of Songs', perhaps the most extraordinary love poem – or, more plausibly, series of love poems – ever written, included in the Christian Bible no doubt because it can be interpreted (rather doubtfully, to be honest) as being addressed to God:

> I opened to my beloved,
> But my beloved had turned and gone.
> My soul failed me when he spoke.
> I sought him, but found him not;
> I called to him, but he gave no answer.

Charisma, sex and religion

It seems that this sense of ecstatic religious love can spill over terribly easily into something altogether less other-worldly – explicit sex. For all its metaphysical and moral connotations, religion in general seems to have a surprisingly intimate relationship with sex. At least according to folk wisdom, sexual acts have a special place in the rituals of many pagan and not-so-pagan religions. In the European tradition, for example, sex has a notorious association with witchcraft (intercourse with the devil traditionally being one of the identifying hallmarks of a witch); in the more modern revivals of ancient European pagan religions, ceremonies are sometimes carried out naked and end with the celebrants coupling freely (or at least, so the more lascivious tabloids would have us believe).

Given this, it is perhaps no surprise that sex has slipped into the edges of mainstream religion with monotonous frequency. In late medieval Europe, the Brethren of the

Free Spirit, Klaus Ludwig's *Chriesterung* movement and the Münster Anabaptists all advocated free love. Each of these attracted tens of thousands of followers in their respective heydays, to the point of being considered a serious threat by both secular and ecclesiastical authorities. Despite mainstream Christianity's attempts to wipe them out, the Anabaptists and the Free Spiriters survived as cults for several centuries, eventually giving rise to numerous descendents – though it must be said that not all of these now advocate sex as the route to heavenly bliss. The Anabaptists, for example, gave rise to several morally upright sects, not least among whom are the contemporary Mennonites, Amish and Hutterites. On the other hand, some of the descendents of the Free Spirit movement took the sect's ancestral doctrines to heart. Among these were the Ranters, who were prominent in England during the Cromwellian Commonwealth of the 1650s; they provide clear evidence from their own writings that the ubiquitous accusations of promiscuity were not entirely propaganda by their opponents.

In Orthodox Russia, it was said that, even as late as the nineteenth century, the rituals of the Khlysty sect involved celebrants dancing themselves into a frenzy round a fire or tub of water chanting hymns until, in a state of ecstasy and near-collapse, they coupled freely. Hindu temples are, of course, notoriously decorated with copulating couples, sometimes intertwined in quite the most imaginative of poses – a factor that no doubt contributed to the enthusiasm with which some Western youth joined Indian religious communes during the 1960s.

Perhaps, on reflection, we ought not to be too surprised by the fact that sex spills over into religion. After all, sex spills over into almost every aspect of human life, especially in those areas that involve an intensity of experience. The surge of endorphins and other neurochemicals associated with ecstatic experiences seems to make us feel especially warm and responsive towards those with whom we are doing the relevant activity, such that only the most stern of religious and moral strictures will suffice to keep the lid on our emotions and us on the straight and narrow. But in the absence of such controls, matters can easily spiral out of control. There you are, coolly minding your own business, when suddenly your hormones take over and flip you into a completely different state of mind. How often have you said: 'I didn't really intend to, but . . .'? Against your better judgement, the neuroendocrines have lifted you up and thrown you helter-skeltering uncontrollably down the slope into surrender.

Of more interest to our immediate concerns, however, is the role that the charisma and sexual attractiveness of a movement's founder play in its success. Women featured prominently in the entourages of many of the medieval Christian sects, including those of Tanchelm of Antwerp and the Free Spiriters. Jan Brockelson, the last leader of the Münster Anabaptist community before it was destroyed in 1535, had to shoulder the holy burden of servicing no fewer than fifteen women towards the end of his career (and that didn't include the wife he left behind in the Low Countries).

Although the more established Christian churches have, of course, been at pains to discourage sexual antics of this kind, there have been some spectacular examples of individual members of mainstream churches who have indulged this particular aspect of their religious vocation with particular enthusiasm. Not least among these were the Renaissance popes, the most infamous of whom was unquestionably the Borgia pope Alexander VI – widely credited with having fathered a child by his own daughter, as well as having many other illegitimate offspring, one of whom he appointed a cardinal at the tender age of eighteen (the boy had already been bishop of Pamplona for three years by then). In more modern times, we have become used to the steady string of TV evangelists whose peccadilloes have eventually been their undoing: Billy James Hargis, Jim Bakker, Jimmy Swaggart, Earl Paulk and Tony Alamo all became memorable for quite the wrong reasons.

But there are also many historical examples that were just as infamous in their own times. In Victorian England, the Reverend Henry Prince's sermons attracted large crowds of adoring female disciples, until the disapproval of his bishop led him first to declare himself the Prophet Elijah (he later reconsidered the wisdom of this and elevated himself to God instead . . .) and then to abandon the stuffiness of the Victorian Church of England altogether in favour of setting up his own religious emporium of some sixty followers – mostly women, whom he christened, with stunning lack of irony, the 'Brides of the Lord'. In the dying days of Tsarist Russia, the 'Mad Monk' Rasputin acquired a reputation both for his capacities as a healer and for his influence over women, not least the Tsarina Alexandra and the ladies of the court. There were dark tales of sexual innuendo.

The leaders of breakaway sects have typically been less than coy about these things. John Humphrey Noyes was not known as 'Father Noyes' for nothing within the

Oneida Community that he founded in 1848 in upstate New York: after the age of fifty-eight, he fathered at least eight children. In the early part of the twentieth century, Joshua the Second (he was originally christened Franz Creffield), Krishna Ventra (formerly known as Francis Pencovic) and Brother Twelfth (his baptismal entry reads, rather prosaically, Edward Arthur Wilson) all applied themselves earnestly to God's work among their willing female disciples. Famously, on 17 July 1831 Joseph Smith, the founder of the Church of Jesus Christ of Latter-Day Saints (better known as the Mormons), had a most convenient revelation from God instructing him that men should marry polygamously – or as God put it rather more delicately, engage in 'plural marriage'. It is said that the revelation came about because his wife Emma was less than enthusiastic about his suggestion one evening that he should take on a second wife. Luckily, God came to his rescue and issued an edict, and she could hardly gainsay it once God became involved. Smith duly took his duties very seriously and allegedly married thirty women ranging in age from fourteen to forty, no fewer than a third of whom were already married to other men.

Continuing this hallowed tradition into modern times, David Koresh of the Branch Davidians claimed to have fathered at least twenty-one children in the community, mostly, it is said, by virtue of claiming the right to sleep with each and every woman in his church, whether or not they were married to another member of the congregation. Astonishingly, the husbands seem to have accepted this practice as enthusiastically as their wives. Then there is the justifiably infamous Bhagwan Shree Rajneesh, dubbed the 'sex guru', of the Oregon Rajneeshpuram. He was said to have been so charismatic (at least in his younger days) that many of his women followers fell in love with him just hearing him speak.

Why are women so especially drawn to these individuals and why do these men behave in this way? The easy answer might be that women are more religious than men, and so more easily influenced. That is certainly a truism, but it doesn't really explain why women should be willing to do more than just worship. One answer might lie in something much more fundamental: the fact that, as we saw in Chapter 4, status – and the wealth that is usually closely associated with status – is an important (though by no means the only) factor in women's mate-choice strategies. I am not, of course, suggesting that religion arose as a by-product of human courtship. Nor am I suggesting that the only thing that motivates men's interest in religion is the reproductive benefits they might obtain thereby. But I am suggesting that religion has often been exploited for nakedly sexual purposes. And it has been so because men have discovered that it is a strategy that seems to work with remarkable ease. It looks suspiciously like another example of the work of sexual selection – able to exploit phenomena with remarkable ease whenever there is some trait that enables judgements to be made about different individuals' qualities as mates.

One question this raises is whether men have been able to exploit this particular loophole because women are more religiously minded than men are, and are thus more likely to be swayed by pre-eminence in this particular arena. Folk wisdom suggests that, despite the fact that men outnumber women as priests and religious authority figures, women are the mainstay of congregations in almost all religions, and this is backed up by research suggesting that women are more likely than men to be involved in religious activities. It's conspicuous that as religion has ceased to be a route to preferment (as has been the case for most mainline religions in the West), so men have lost interest in careers as religious specialists. Vocations from men for the mainline Christian churches have dropped like a stone in the past half-century, triggering an increasing acceptance of women being allowed to train for the priesthood. Perhaps not surprisingly, men still seem to dominate the stage in the more charismatic sects, however.

One reason for women's greater interest in religion might be their greater emotional and social sensitivity. Women express emotion more intensely than men, not only in terms of self-report, but also in terms of facial expression and physiological responses (e.g. galvanic skin conductance), and especially so in the expression of negative emotions like sadness and disgust. It has often been claimed that women's greater emotional reactiveness is a product of how they are socialised, and while this may well be true up to a point, there are good reasons for believing that socialisation merely exaggerates existing biological differences. Recent brain-scan studies have shown that men and women respond differently to emotional cues in terms of which parts of the brain are active, and this ties in with evidence that women respond more deeply than men to emotional events. For example, although the brains of both men and women respond with bursts of activity when thinking about sad events, women's responses are many times more intense than men's responses and tend to involve both hemispheres of the brain more completely. Women also score significantly better than men on advanced mentalising (theory of mind) tests. One consequence of this may be that women are both more sensitive than men to reflective anxieties about the vagaries of everyday life and the uncertainties of the future, and better able to reflect on other people's mind states. As a result, women may be more prone to seek structures and processes like religion that bolster their ability to cope with these anxieties. Given the role that religion plays in this respect, these effects would tend to make women more susceptible to religious persuasion.

A second predisposing factor is probably the fact that charisma is a major issue in women's mate-searching strategies. Charismatic individuals, whether political or religious leaders, sports stars, musicians or even occasionally writers, seem to be especially attractive, and notoriously so in the case of male pop icons who continue to attract more than their fair share of free sex. The reasons for their attractiveness might be either something to do with their status (and hence potential wealth) or their good genes (as implied by their intellectual or physical skills). So powerful is this effect that it seems to generalise even to virtual individuals such as God. This extension of the natural processes of being attracted to someone, or falling in love, to virtual individuals may be a consequence of the fact that when we do this in everyday life we are in fact constructing a vision of the person in our minds. In other words, we are falling in love with an image we have constructed that is only partly informed by what is actually in front of us. I'll say more about this in the next chapter when I consider falling in love on the Internet.

A third factor may be women's greater tendency to

commit themselves to a prospective mate. Although folk wisdom suggests that human courtship involves ardent males courting coy choosy females, the reality is that the business of choosing mates is a two-way process. Both sexes have a great deal at stake, even though their strategies for achieving the desired end (netting the best partner available) may be quite different, as may be the ends that they seek to achieve. Men do court and women do choose (and women's decisions are probably made more carefully than men's, as Chapter 4 suggested). Nonetheless, women can become very committed and persistent in their pursuit of individual males once they have made their minds up. Men are typically not quite as focused in their courtship strategies, and give up more easily if they feel they are unlikely to succeed.

The God spot

Neuroimaging has begun to provide some insights into the processes involved. A recent neuroimaging study revealed that when people are thinking about God or other spiritual beings, the same areas of brain are active as are involved in mentalising (see Chapter 3). Clearly, we see spiritual beings as real people. Some years ago, Andrew Newberg (a neuroscientist) and Eugene d'Aquili (an anthropologist) found that individuals in the state of religious ecstasy produced during meditation have a unique pattern of brain activation. They show greatly reduced levels of activity in the left posterior parietal lobe (in effect, near-complete shutdown) and a great deal of generalised activity in the right hemisphere, usually associated with more unconscious, emotional responses. The area of the parietal lobe that appears to shut down during mystical states is known to be implicated in our sense of spatial self, and this almost certainly accounts for the sense of detachment from the real world that is associated with states of ecstasy. Newberg and d'Aquili argued that when the neurons in the parietal lobe start to shut down, they release a series of impulses via the limbic system (the amygdala again) to the hypothalamus, which sets up a feedback loop between itself, the attention areas in the frontal cortex (which are normally responsible for inhibiting the parietal lobe neuron bundles) and the parietal lobe. As this cycle builds, it leads to the complete shutdown of the spatial awareness neural circuits, generating as it does so the burst of ecstatic liberation in which we seem to be united with the Infinity of Being that is so characteristic of entering a trance state. They labelled this bundle of neurones in the parietal lobe the 'God spot'.

The fact that the hypothalamus is involved rather suggests that endorphins might also be implicated: this is, after all, the main location where endorphins are produced. More importantly, many religious practices (and especially those involved in effecting trance states) involve pain and physical stress on the body. In some cases, it may involve the deliberate infliction of pain. The most infamous case in Christianity is, of course, the medieval flagellants who toured Europe during the Black Death, whipping themselves into a frenzy and attracting huge crowds in the process. Penitential discipline has a long tradition in Catholic monasticism, for both men and women, where 'the discipline' referred to the knotted cord cat-o'-nine-tails traditionally used during private prayer. In the Russian Orthodox tradition, the above-mentioned

Khlysty ('flagellators') and the Skoptzy ('mutilators') sects aimed to achieve a state of religious ecstasy through self-imposed pain. (Since Skoptzy practices involved such things as slicing off women's breasts, they had, needless to say, a relatively short-lived popularity. The more moderate Khlysty sect, however, had a long history from as early as the 1360s until as late as the 1890s.)

These practices are by no means confined to Christendom, of course. Islam has its own versions. The so-called 'whirling dervishes' of the Mevlevi order, which dates back to AD 1273, dance themselves into an ecstatic frenzy using a mesmeric gentle twirling motion. More extreme are the 'howling dervishes' of the Refa'i order, founded by a descendent of the Prophet Muhammad's grandson, Hussein, who slash at themselves with knives or pierce their bodies. These are best known for the annual Shia ceremony of Ashura that celebrates Hussein's martyrdom at Karbala in modern-day Iraq in October of AD 680. The massed ranks of penitents process through the town to his shrine, whipping themselves over their shoulders to the point of drawing blood.

Other recent neuroimaging studies suggest that religious belief states (and especially beliefs about God's surprisingly emotional responses to our behaviour) engage many of the same brain units involved in theory of mind. An important finding for present purposes is that religious *knowledge* and religious *experience* seem to activate different areas of the brain. Religious knowledge activates areas in the temporal lobe (adjacent to the ears), whereas religious experience activates areas in the left frontal lobe and the parietal and temporal lobes. The first probably reflects the co-ordination of associations between intellectual facts, whereas the second might be more closely related to the areas that underpin theory of mind and mentalising.

Beyond the edge of despair

As a generality that brooks almost no exceptions, it seems that this falling in love business is always directed at one individual. We seem to find it genuinely difficult to fall in love with two people at the same time. In some cases, this focus on one person can, however, be taken to extremes, producing a rare but well recognised condition known as de Clérambault's syndrome, after the French psychiatrist Gaëtan Gatian de Clérambault who first described it in 1921. The key symptom of this syndrome, which is about twice as common in women as in men, is that the sufferer has a delusional belief that a particular individual is very much in love with them. This belief is usually so intense that nothing the object of desire says or does will shake it: all attempts to put a stop to the unwelcome attentions of the individual concerned are interpreted as playing hard to get in order to test the suitor's ardour and resolve. Being rude or angry with the sufferer simply results in them redoubling their efforts, now even more convinced that their initial beliefs about your interest were right. Although this condition is often associated with other psychopathologies earlier in life, it may in fact be just the tip of an iceberg, a reflection of women's more focused courtship strategies taken to extreme – much as autism is the floating tip of a broader male cognitive style that is naturally less intensely social.

There are some striking sex differences in the incidence

of de Clérambault's syndrome. In men, it typically involves younger, lower-class individuals, and something rather similar can be seen as a form of exaggerated jealousy in younger men who have been abandoned by a romantic partner. In contrast, it tends to occur in older women, and more often women who are or have been married. Men tend to focus their attention on women who are younger than themselves, whereas women typically focus on men who are older or of higher status.

De Clérambault's syndrome has become a serious problem for celebrities. Among the best-known real-life cases was Margaret Mary Ray's obsession first with TV host David Letterman and later with the Skylab astronaut Story Musgrave. Ronald Reagan had an unfortunate encounter with the syndrome that nearly cost him his life when John Hinckley Jr tried to assassinate him in a desperate attempt to impress the actress Jodie Foster, whom he had been stalking. Tatiana Tarasoff was less fortunate when her fellow student Prosenjit Poddar stabbed her in an attempt to engineer a situation in which he could seem to rescue her from danger; unfortunately, he misjudged things and killed her. The syndrome has also featured in literature. It is central to the plot of Ian McEwan's novel *Enduring Love*, though, unusually, in this case the sufferer and the object of desire are both men. And it has featured in at least three episodes of the CBS *Criminal Minds* police drama series. And one cannot help feeling, when reading her writings, that St Thérèse of Lisieux at the very least showed tinges of it.

From an evolutionary point of view, this kind of obsession makes a certain sense. Once you have made up your mind whom to take as your lifelong partner, then the best thing is to go for it for all you're worth. He, or she, who hesitates is lost. Make your intentions as clear as a bell, and keep battering away . . . and, eventually, even the biggest dullard will get the message. Often as not, it's just the message they need to get: once they realise that you are interested in them, it's usually all they need to go through that magical instantaneous flip and suddenly capitulate. Persistence is a strategy that often pays off in the mating game. It's easy to see some advantages to a psychological mindset that in essence says: 'If you see the perfect partner, just go for him/her before someone else does.' Although women are impressed by attentiveness, men are probably more easily swayed by persistence – if only because they are usually happier than women to settle for whatever they can get.

In many ways, the problem is created by a combination of circumstances. Life is much too short to waste time searching for Mr or Ms Perfect. There is a fundamental trade-off between continuing to search forever and just getting on with the biological business that all this is about – reproduction. At one level, it probably doesn't matter *too* much whom you do your reproducing with, providing they have a half-decent set of genes and aren't infertile. You probably won't do significantly better by rejecting an endless series of suitors in the search for the Perfect Mate. While there may be an advantage in not settling for the first prospect you happen to come across, there are diminishing returns to be gained by extending your search for too long. There was a famous set of theoretical analyses in the 1970s which suggested that the optimal strategy was to check out the first thirteen prospective mates you came across and then take the best of the bunch. You were unlikely to do better. Later, Peter Todd and Geoff Miller ran a series of computer simulations and proposed the 'Try a Dozen' strategy: check out the first dozen prospective mates you come across, decide which of those is the best, then wait for someone new to come along who outshines them, and go for that one for all you're worth. Since life is fixed at the other end, delaying longer becomes an increasingly bad option.

There are two reasons why this might be so. One is that the business of mate searching is, as I observed in Chapter 4, a frequency-dependent problem. Even if the Perfect Mate really does exist, they know how perfect they are (if only because everyone is chasing them) and so will only be willing to settle for the Perfect Mate of the opposite sex. They end up with the pick of the bunch. Mr Perfect waits around until he comes across Ms Perfect. Once they have paired up, Mr Next-Most-Perfect settles for Ms Next-Most-Perfect, and so on down the line until it finally gets to you. But since you have held out all this time hoping that Mr or Ms Perfect might be persuaded by your overtures, you have wasted an inordinate amount of time before finally coming to the realisation that you are going to have to settle for what you can get anyway. Meanwhile, your biological clock has been ticking relentlessly, and your reproductive opportunities are dripping away before your eyes like the proverbial sand in the hourglass. In other words, we are biologically programmed to check out only the local corner of the market, and then just get on with it, for a very good reason.

A second reason why this is inevitable is that there is in fact no such thing as the Perfect Mate. They simply don't exist. Sadly, none of us is *so* perfect, if only because there are too many dimensions to perfection that are often in fundamental opposition, so that perfection on one dimension can never be satisfied with perfection on another. Psychologically, however, you need to believe they exist so that, once you have made your mind up, you can commit everything to that one person, otherwise you won't have the staying power to stick with the one you finally decide to settle for – especially once the rosy spectacles of first love fade. Even if you only remain with each one for a brief period before moving on to a new mate, you have to believe that each successive mate is Mr or Ms Perfect. Once you come to realise that this is just an illusion, your lack of commitment and disinterest will be so obvious that only the most besotted will fail to see through you.

*

Religious love is safe love. There is little risk that the object of desire will betray you. Like the troubadours of medieval Europe, you can worship your beloved from a distance without the complications and inconveniences of everyday relationships. But even if he (or she) does occasionally let you down, a mild version of de Clérambault's syndrome allows you to retreat into the belief that your love is simply being tested. This is particularly characteristic of the Abrahamic religions. The Bible is awash with instances in which individuals' resolve and commitment to God is tested, sometimes in the most extreme ways. Even if God does not himself do it, he allows Satan to do it for him. Those who falter lose the race; those who stand firm in their belief in the Beloved eventually triumph.

But there is another sense in which these relationships are safe. Because it is all in the mind, you can invent the perfect partner. Your dreams can never be contradicted by the intrusion of brute reality. The Beloved is tailor-made for you because you make it so. There need be no blemishes of character or form, because you can construct the Beloved to mirror precisely the traits you long to have in the perfect partner. In the next chapter, we will meet another case of this in the form of virtual reality – love on the Internet.

9

Love and Betrayal Online

Wi' lightsome heart I pu'd a rose
Upon its thorny tree;
But my fause luver staw my rose,
And left the thorn wi' me.[*]
'The Banks o' Doon'

We have been lucky enough to witness one of those extraordinary moments in history that set in motion a major change in our lives at a single stroke. As we slipped quietly into the new millennium, our world acquired a new dimension: the Internet. Facebook, more than anything else since the invention of the postal service, has revolutionised how we relate to each other. Finding love and fulfilment has never been so easy: the availability of Internet dating sites, those emails from friendly Nigerian spammers offering untold riches in exchange for the details of your bank account, and promises of eternal love from Russian maidens have all widened our horizons in the way that neither the penny post of the 1850s nor the telephone of the 1930s could ever possibly have done. At a keystroke, we became members of the global village.

In the digital world, the dream was, and still is, that new electronic ways of communicating promise wonderful new vistas, a new Wild West to be explored and won, and all from the comfort of your own home. The miracle of the Internet will provide us with connections so vast and pervasive that our social lives will be enriched with new

[*] Pu'd: pulled. Fause: false. Staw: stole.

friendships across the globe. The Gordian knot created by the limitations of face-to-face interaction that has, until now, bound us to our small urban worlds – the handful of people that we meet at work and at play in our everyday lives – will be cut through. We can broadcast, literally, to the world.

In one sense, this is all true. However, the spanner in the works that undermines the upbeat claims of the technologists desperate to sell their software is the human mind. The techies have forgotten that the constraint on the number of friends we can have is not just a technicality, or due to the time we have available in our busy lives: it also has something to do with whether we can manage to maintain coherent relationships with more than a handful of people. Yes, time is a constraint, and the digital world offers us a way around that because we can speak to several people at once. But the real issue is the subtle distinction between speaking and talking. We can speak to the world, like a lighthouse endlessly flashing out its message across the sea to each and every passing ship, but how many of those passing ships can we hold meaningful conversations with? And, eventually when the boats do draw close, can we genuinely go on board more than one at a time for the social events on offer?

Mirage and avatar

In many ways, the secret behind Facebook's success has been that it allows us to keep up those relationships we value over distances that, in the past, would have meant that the relationship died a natural death within a relatively short space of time. Until a century or so ago, our worlds were small and confined: relatively few people moved long distances, and those who did more or less lost contact with their natal communities. But in our modern mobile society, we move at regular intervals for education and jobs. And when we do, we gather and leave behind small clusters of friends. Our networks become fragmented so that they consist of several discrete sub-networks which do not overlap that much. Our home-town childhood friends might know our parents and siblings, but they won't know our extended kinship circle, and our friends from college days won't know either; later, when we have built and left two or three more clusters of friends in different cities as we moved jobs, we add successive groups whose sole acquaintance with each other may only be a passing remark on Facebook, or a brief encounter at one of those once-in-a-lifetime gatherings to which we invite everyone we know – usually a birthday with a zero on the end.

In one sense, that fragmentation leaves us vulnerable. Our social world in any given place is small, and we can become isolated more easily. If we lose a close friend for some reason, the ramifications are more intrusive because we have nowhere else to retreat to, no extended family or network of friends and friends of friends on whom we can rely to fill the gap. Modern life is isolating and more liable to social exclusion than any we have known in our evolutionary history.

The Internet provides a solution because it allows us to maintain contact with friends who are scattered over continental spaces. As a device for keeping friendships alive, it has enormous potential and has been used with considerable effect. It also has the great advantage of allowing us to speak – at least metaphorically – to several, perhaps even many, people at the same time. It allows us, in principle, to build extended networks of people whom we have nominally 'friended', even though in reality we know rather little about them.

Inevitably, of course, the Internet has played its role in romance, too. Finding suitable partners has never been easy, especially once populations became more mobile and our social world became increasingly based around those with whom we work. In traditional societies, marriages were contracted on the community scale. Everyone had a view. It mattered to the community as a whole who settled down with whom. Advice was sought, and freely given even when it wasn't sought. Through the extended networks, proposals were offered, suggestions made. And in doing so, our networks offered surety and stood guarantee. But we have little of that luxury in the modern world. In our dispersed and fragmented social worlds today, we lack those opportunities and sureties.

Beginning with the increasing urbanisation at the end of the eighteenth century and exacerbated by the industrial revolution of the nineteenth century, social and economic mobility has removed many of the traditional means by which we gained access to mates. The old village matchmaker could not operate outside her local community. We were on our own. And as the nineteenth century progressed, the problem just got worse: it was ramped up by the dramatic rise in the number of women working as domestic servants far from home in the grand houses of the rich during the second half of the century. It led inevitably to the rise of professional matchmaking services. The first matchmaking bureau opened its doors in London in 1751, offering to connect strangers of suitably genteel quality.

However, it was in the late Victorian period that these services were to blossom into a veritable industry. During the 1880s and 1890s, personal advertisements became immensely popular, and not just in the remoter corners of the New World and the Antipodes where women were in short supply. They were popular even in Victorian and Edwardian London. The *Matrimonial Post, Matrimonial Herald, Matrimonial Record, Fashionable Marriage Advertiser, International Matrimonial Gazette, Matchmaker, Matrimonial Circle* and *Matrimonial Times* all survived until the First World War. What killed the entire industry was not the war itself, but the fact that it gradually became associated with the seedier elements of the Edwardian sex trade. Prostitutes came to realise that personal columns offered a perfect venue for advertising their various services under the guise of matrimonial offers. Genteel women ceased to advertise for fear of receiving the kinds of replies they had not envisaged.

It was not until after the 1960s that there was a resurgence in the personal ads business, when increasing economic mobility meant once again that many found themselves in situations where they could not easily meet suitable members of the opposite sex. That trend has simply continued into the digital age with online dating services replacing hard-copy ones almost entirely by the millennium – at which point they were joined by innumerable specialist sites offering Russian and Indonesian girls looking for matrimonial bliss in the West. The Internet has merely provided a fast, worldwide medium for old-fashioned lonely hearts columns. Most operate in much the same way as their predecessors, asking users to profile themselves and matching their profiles to appropriate individuals on their database – or by allowing punters to contact the advertisers they might be interested in directly. Computers simply make the processing faster and more comprehensive. Many of these sites now exploit the findings of psychological, and especially evolutionary psychological, research of the kind we explored in Chapter 4. In most cases, users are asked to pay a fee on signing up and some indication of the sector's popularity and success is provided by the fact that, between them, these sites are estimated to have a billion-dollar turnover. In fact, according to one survey, 15 per cent of all those who use the Internet at all do so to meet people with a view to forming a relationship and finding romantic fulfilment.

Virtual love

The great difficulty with the new wave of Internet dating sites is that one can never be quite sure who is actually behind the profile or the email. We are back in the uncertain world of Edwardian matrimonial magazines when, unbeknownst to us, all kinds of dodgy deals are being done. This is not a peculiarity of the dating websites, but rather of Internet culture as a whole. It is only too easy to create a persona for oneself. And so it is that teenage Nigerian boys sitting at the only Internet connection in their village can fill our inboxes with often illiterate offers of untold riches – if only we will tell them our bank details – or heartfelt pleas of eternal love in exchange for . . . well, our bank details, or a modest cheque to help them pay for desperately needed medical treatment. Without – and sometimes even with – the opportunity to see a photograph of the person at the other end, we have no idea whether the sweet young thing who will die of some dread disease if we don't befriend them immediately is actually a sweet young thing or a hoary old man, or a teenage wannabe-rapper with too much spare time on his hands. Everything, including the carefully posed photographs, can be invented. It's the murky world of stalkers, only now they want more than just your body. What should – and would – have worked just fine has been undermined by freeriders intent on exploiting our foibles and weaknesses for their own benefit. The loophole they exploit is our desperation, often fuelled by loneliness and the looming prospect of an even lonelier old age, combined with our natural predisposition to trust, to take everyone at face value.

And herein lies the essence of the problem. When our interest is roused by an initial exchange, we invent the person we are falling in love with even in real life. It's the rosy sunglasses of romance that are needed to overcome our scruples and social reserve and give us the courage to get started. A constant theme of the sorry tales of woe that come from victims of Internet fraud is the speed with which they moved from healthy scepticism to utterly committed, unstinting, desperate love for the unknown (or sometimes, even known) person at the other end of the email or phone call. In case after case, it seems to happen almost overnight, and often only a few weeks or months into the exchange of messages.

In real life, we meet the person every day and gradually accommodate our mental image to reality. In the digital world, we don't. So when we fall in love on the Internet, we fall in love with a mental image we have constructed for ourselves, not with a real person. And we don't have the anchor of reality to hold our wild imagination in check. We simply get drawn ever inwards into the centre of the spider's web.

It is not known for sure just how many people are deceived by spammers and fraudsters on the Internet, or how much money is lost. Most people are probably too embarrassed to report it to the police when it happens. Nonetheless, the most conservative estimates suggest that Internet fraud accounts for billions of dollars each year. And a significant proportion of this is the result of victims willingly handing over substantial sums of money in the mistaken belief that the person at the other end is deeply in love with them and just needs a helping hand to fix their life so that they can then buy the crucial plane ticket to come over to live with them in bliss and harmony all the remaining days of their life. . . . Many of these victims are lonely single women in search of a little TLC and attention who can ill afford to be taken advantage of in this way, either financially or psychologically.

In one recent case, fifty-two-year-old Grenadian-born David Checkley was jailed for six and a half years for fleecing some thirty women in the UK – most of whom he met through Internet dating sites – out of around half a million pounds between them. Some lost their businesses, several lost their houses, most lost their life savings, and all lost their dignity and had their trust in people shattered. He had sweet-talked his way into their hearts, exploiting their gratitude that someone – anyone – was taking a serious interest in them, offering friendship and love. Ever the attentive gentleman, his behaviour at first seemed chivalrous and generous. Then he needed a quick loan of a few hundred pounds to tide him over a business deal. It was invariably given willingly. And then another rather bigger loan was needed, and then another. And always some reason why the earlier loans couldn't be repaid just at that moment.

Romantic spamming is a psychological art, often perpetrated by people who are not especially well educated. But they understand the victims' psychology and weaknesses and can exploit the loopholes in their defences in what is invariably a masterclass in deception. Yet it is still puzzling that the victims are taken in by claims which often, on investigation, turn out to be – and are then often freely admitted by the guilty party as being – lies. Usually, some reason is given to explain why the fraudster needed to adopt a persona, or to claim to have business or charitable interests that in fact were non-existent. Here's a very typical sequence of events based on real-life cases:

The victim, a well-educated, professional woman, perhaps in her fifties with several grown-up children, is on the rebound from a failed marriage or, perhaps, is growing increasing lonely after being widowed. The prospect of growing old alone looms larger by the day, and she is desperate to avoid so unhappy a fate. So she sets about looking for a partner, a companion, often with no explicit sexual interest, through an online dating service. It seems safe – after all, users have to pay and that has to be a deterrent to casual predators. Face-to-face meetings are arranged with what turn out to be some hopelessly unsuitable men. All very disappointing. Then she receives a tentative contact that is different from the usual run-of-the-mill responses to her posting. Intrigued, she answers. The name is exotic, the photo when it eventually comes invariably shows a Mediterranean god. Yes, in retrospect, he was vague about his age, and, yes, he was vague about a lot of things, and the name he gave should have raised alarm bells because, had she thought about it, it didn't really sound like the kind of name one would associate with Greece/Italy/Spain or wherever he had said he came from. But she was getting desperate.

A few more email exchanges, then a phone call. He doesn't sound southern European at all. In fact, his English is so poor and riddled with patois that he could only be West African. So she challenges him, and he laughs and freely confesses to the subterfuge – with some plausible cover story: he didn't want to put her off before she had even had a chance to see the real him and how nice he actually is. She knows that strategy, and has probably used it to good effect herself, so is willing to forgive. A wry smile, but after all it's a small matter in the grand scheme of things and he seems gentle and generous . . . and loving. A few more email exchanges, and then a shy declaration of love. Despite all the little deceits, she is already convincing herself that she is in love too. He would like to come and be with her, but he doesn't have the money. A business arrangement is casually mentioned, only it needs some seed money to get it off the ground. It will pay a handsome profit, and then he will be able to buy that plane ticket and be with her.

At some point, she makes the mistake of casually mentioning – as one does innocently enough in conversations with those one trusts – that she recently inherited a sizeable amount of money. One thing leads to another, and a trip is planned to meet up in a safe haven – all paid for, of course, by the victim. A few weeks of heavenly bliss. She feels loved for the first time in many years. All the lies and deceits are forgotten and forgiven. When confronted with reality, we are always willing to compromise on our overly picky standards in order to avoid the Cinderella moment of going home alone after the ball. It's Jane Austen's curate option, Hobson's Choice, the best of a bad job. Convinced now that she has finally glimpsed nirvana, she lets go and allows the stream to carry her whither it will. Then the big business deal is proposed. It just needs a big loan for a few months. It's as safe as houses. She hesitates, so emotional blackmail is laid on thick: how can you deny me so little a thing when you love me so much? She capitulates against her better judgement, suppressing the nagging doubts and the alarm bells ringing in the deep recesses of her mind. They part, he to follow up on the business deal, she back home to wire the money and prepare the nest.

And then comes the let-down. The phone call, or more likely email, to say that the business deal has gone belly up, the money is lost. So desperately sorry . . . a terrible thing . . . nothing to be done . . . one of those things . . . what must you think of me . . . what can I do to make amends . . . and so on and so on. But by then it's too late. She is in over her head and the waters are closing fast. She can back off and lose everything – money, love, self-esteem – or she can take the charitable option and be sucked in further still in the hope of salvaging something for the future, only to be exploited yet more.

Women seem to be especially prone to these romance scams, but men can just as easily fall prey to the pretty young thing who just wants to find the perfect husband, the Prince Charming to whom she can devote herself. The only difference is that she mostly does have to be a young thing for the entrapment to work. And if it's not a financial scam, it's a scam to obtain a UK passport on marriage – followed swiftly by a disappearing act or a simple request for a divorce. In the end, it's all a form of highway robbery, carried out on the twenty-first-century virtual equivalent of Dick Turpin's trunk road to York. And like Turpin in his face mask and ankle-length riding cape, the scammer keeps his or her identity well hidden. Like all the best magic shows, the scam works best when the scammer gives away least, and allows the victim to invent the story. When you come to believe something for yourself, you believe it with greater conviction than when someone else tries to convince you.

Cyberwars

Broadly speaking, the world consists of two types of people: cyberoptimists and cyberpessimists – those who think the Internet is the answer to all our problems and those who are very sceptical of such claims. It seems that there are no intermediate positions. Cyberpessimists take the view that the Internet is having a negative effect on our social life because time spent online is time we don't spend in social engagement with real-world friends. There is a certain logic to this, especially for children and teenagers, who according to most statistics are spending increasing numbers of hours online each day. The skills we need to steer a course through the complexities of the adult social world take decades to learn and, like all learning, require real-life experience. We have to learn how to accommodate others, how to tweak what we want with the needs and desires of those with whom we interact. It is less easy to do that on the Internet, because you can simply pull the plug and fade away in a way that is not possible face to face without creating more direct offence. More importantly, we need to learn the skills of interpreting the subtle facial signals of other people and how to use these to adjust our own responses. That can only be done face to face with the real thing, and it takes time to learn.

Learning the fine-grained meaning of facial cues and gestures is a long slow process. In fact, it has turned out to be much longer and slower than we might ever have imagined. Quinton Deeley, one of my collaborators, scanned the brains of some forty males aged between eight and fifty years old while they were doing a facial cue recognition task. The task was very simple. Subjects were shown a series of photographs depicting fearful, disgusted or neutral faces and asked to identify the emotion being expressed. The parts of the brain that were differentially active when they were processing the task were the usual suspects that one might expect: the visual system, the face-recognition neurons in the fusiform gyrus of the temporal lobe and parts of the right frontal cortex that are known to be involved in emotional responses. Nothing hugely novel there.

What was unexpected was that the level of activity in the frontal cortex showed a distinct gradation across age. It was very active in younger subjects, but in older subjects activity in the frontal areas dropped to baseline levels and was replaced by an increase (at least in response to disgust emotions) in the fusiform area. It seems as though when we are young we really have to work hard at recognising facial cues of emotion, and so all the work is being done in the smart bits of the brain in the frontal lobes. But as we gain experience, this gradually becomes automated and moved down to other parts of the brain that can deal with it more efficiently offline. What was particularly surprising was the age at which this happened: it was not until the mid-twenties that the switch took place. That's interesting in itself, because that's the age at which the brain finally settles down to its full adult size and configuration. But the real issue is just how late this happens: it suggests that it takes us a very long time and a great deal of experience in real life to reach the point of being well versed enough to automate the response.

We have to learn how to navigate our way around the real social world, how to cope with its vagaries and its constant turmoil. It takes a *very* long time. The acquisition of social skills like theory of mind occurs well after children have acquired all the simpler kinds of everyday understanding about the conservation of volume and causal reasoning. It's not just a matter of recognising a pattern in the visual cues offered by a face and identifying the emotion being registered – as we might learn a particular lighthouse's pattern of short and long flashes. We also have to learn subtle differences in the depth of emotional expression coded in a grimace, and the difference between true and fake emotions. We have to learn the cues that differentiate an involuntary, genuine, Duchenne smile from a voluntary, faked, non-Duchenne smile, and the significance of the presence or absence of those tiny crows'-feet creases in the corner of the eye that signal the difference. Until we do, a smile is a smile is a smile, and we will be deceived by the avatar before us.

Learning the subtleties of social cues is a reason for not becoming too heavily engaged with the Internet when you are young. But there are other reasons why cyberpessimists remain suspicious. One of these is the fact that time is, in the economists' parlance, inelastic. You can't compress it so as to use it more efficiently. This becomes important in the context of how we maintain our relationships. As I explained in Chapter 6, the emotional quality of our friendships (but less so our kinships) is dependent on the time we actually spend interacting with them. The more we do, the more we become emotionally attached to the individuals in question. Time spent on the Internet is time taken away from interacting with family and friends face to face. And it is this that seems to be critical, especially for males. This came out very clearly in our longitudinal study of mobile phone use: what prevented the boys' friendships declining over time after they had moved away from home was time spent *doing stuff* face to face. This wasn't true of the girls in the sample, for whom the critical activity was time spent talking together. And for this, Internet chatrooms and the phone are almost perfect substitutes. That probably explains why around two-thirds of Facebook users are women. Men, on the whole, find it a less congenial environment. However, we suspect that, even for the girls, technology is ultimately only a sticking plaster, a temporary fix, something to tide you over the interim. In the end, if you don't get together for a face-to-face chat, nothing in the digital world will prevent that relationship sliding down the slippery slope into the abyss beyond the 150 relationships that we can typically manage.

We devote around 40 per cent of our total social time to our five closest friends and kin, and around 60 per cent to our fifteen best friends and kin. A simple calculation shows that if we were to devote the same to everyone in our wider social circle, we would rapidly run out of time.

Our social time accounts for around 20 per cent of our eighteen-hour waking day. To devote the 5 per cent of time that we give to each of our fifteen best friends to all 150 members of our full social network would require that we devote twenty-seven hours a day to face-to-face interaction. If we wanted to have all our relationships at the top-gear intimacy of our innermost five relationships, we would have to spend an even more impossible forty-three hours a day in face-to-face interaction. We just couldn't do it. Not only is the day not long enough, but we wouldn't have time to eat, sleep or draw breath.

So perhaps inevitably, despite the cyberoptimists' bullish claims that the Internet offers unparalleled opportunities for opening one's social horizons by providing the possibility of meeting tens of thousands of new friends all around the world, our online social networks turn out to be much the same as our everyday, real-world, offline networks. When Tom Pollet, Sam Roberts and I undertook a study of online social networks, we found that those who used Internet social networking sites like Facebook or other chat sites did not have larger social networks than those who rarely used the Internet. Judith Donath and danah boyd (who always spells her name that way) came to much the same conclusion in their analysis of social networking: social networking sites don't allow you to have more close friends, even if they do allow you to have more casual acquaintances. And the issue here is the word 'allow': just because you *can* have more acquaintances (those who are known in the networking trade as 'weak ties'), it doesn't mean to say that you *have* to. And, in fact, it seems that most people don't: despite all the hype, most of us seem to restrict our Facebook friends to the same group of people we know in real life. When Facebook themselves did a trawl through their enormous database, they found that the typical number of friends was between 120 and 130, comfortably within the limiting value of the 150 limit on our relationships. Yes, some people do have many hundreds, even thousands, of friends on Facebook, but in reality they are few and very far between. Most people want to feel that their Facebook list *is* their everyday social world, even if they do have one or two cyberfriends in addition. The fact is that we just cannot manage more friendships.

Indeed, when they are posting on their Facebook 'wall', most people write as though they are in a real-life conversation: they think they are speaking to just a handful of their best friends. In the real world, the upper limit on the number of people who can maintain a functional conversation is just four: add an extra person, and I can guarantee that within thirty seconds it will be two separate conversations. Check it out next time you are at a big reception. So when we post something on a Facebook wall, we invariably think we are engaged in a private conversation. The rest are simply voyeurs on your conversation, and if they weren't there, it wouldn't matter that much. We had this same impression from our longitudinal study. In our modest sample of eighteen-year-olds, we had three who, over eighteen months, sent an average of over a hundred texts *a day*, and presumably they received the same number of replies. (Work it out for yourself: that's over 108,000 texts sent and received in eighteen months!) But what was really surprising – and this has been borne out by a number of other studies – is that 85 per cent of these texts were sent to just two people: best male and female friends. Our real social world is *very* small. For this reason, some of the newer social networking sites, like Path, restrict the number of 'friends' you can have (to fifty in the case of Path, who used my research to guide them on this). They explicitly aim to create a greater sense of intimacy, thereby allowing you to feel a sense of implicit trust in those with whom you are sharing photographs and personal information.

So what is it about face-to-face interactions that makes so much difference? We are not entirely sure, but what is clear from our research at least is that we are much less content after a digital interaction than after meeting someone face to face, even allowing for the differences in how long the interaction lasts. In the study that Tatiana Vlahovic and Sam Roberts carried out in my research group, they asked forty-three people to complete a daily diary for two weeks recording every interaction they had with their five closest friends and rating them for the happiness they felt afterwards. Face-to-face interactions scored much higher than phone calls, emails, texting, instant messaging and interactions through social networking sites like Facebook. Only Skype exchanges scored as well as face-to-face interactions. The obvious reason is that only Skype comes close to providing the sense of being right there with the person that a face-to-face interaction provides. To be honest, I was surprised at how well Skype did. I hadn't expected it to be rated as highly as face-to-face interactions because it is still so jerky and clunky. It's perhaps a sign that, as the technology improves, the benefits of Skype can only get better. That sense of co-presence, of being in the same room, seems to make a massive difference. It seems that all the subtle auditory cues of intonation and prosody that we get via a phone call just aren't enough on their own, although, as I mentioned in Chapter 2, hearing laughter does make a difference.

But my intuition is that one thing is yet missing. Our measure of our subjects' contentment with an interaction was rather crude, and reflected only the momentary happiness with one particular event. We did not attempt to measure the impact of each exchange on the overall quality of the relationship. My guess is that the surprisingly positive results for Skype paper over some cracks in terms of the longer-term consequences, particularly for intimate relationships, and especially for the most intimate of all relationships: romantic pairbonds. The reason is obvious when you think about it. Facial expressions serve an important function, but in the end intimacy is really dependent on touch. Not to put too fine a point on it, a touch is worth a thousand words. We learn so much more from the way someone touches us than from anything they could possibly say. Until the techies finally manage to crack the problem of virtual touch, digital media will never be as effective as face-to-face interaction, and will never replace it.

Virtual unreality

In the previous chapter, we came across another case of virtual-reality romance, namely falling in love with God. There were the same hallmarks of a self-created icon, a mapping of our ideals onto a shadowy all-purpose framework. There seems to me to be a number of parallels between these two cases that arise from the way our minds are constructed. The differences lie mainly in the fact that one is innocuous and safe because the object of our desire can never actually impinge on our world, whereas the other is fraught with risk and emotional trauma because ultimately there are real people behind the digital avatars.

I would have liked to end this particular chapter on an upbeat note and enthusiastically extol the virtues of the global village. I am not sure that I can. My sense is that the digital world is too fraught with traps for the unwary, and provides too many hiding places for rogues. In real life, we are inclined to trust people unless and until circumstances or their behaviour prove otherwise. We are predisposed to trust those among whom we find ourselves, partly because social life as we know it would collapse if we didn't. And because that has been essential to the survival of small communities – and hence to their members – since time immemorial, natural selection has instilled into us a willingness to trust others and behave prosocially towards them. In the small-scale world of everyday life, the proverbial secondhand car salesman is easily unmasked. We spot the shifty eyes and other cues, and word spreads rapidly around the community. In the much larger and more anonymous world of the Internet, this doesn't happen.

There is one half-exception to the rule: virtual-reality gaming worlds like Second Life and World of Warcraft naturally seem to develop similar codes of behaviour and ways of self-policing, as happens in the real world. Those who break the rules or the unwritten codes of behaviour that a community shares are ostracised and punished. Perhaps this isn't so surprising, as these environments are designed to create small communities who work together to solve a problem. What's more, they tend to naturally evolve just the same kinds of grouping patterns as we see in the everyday world – small cliques of good friends who grow to trust each other enough to collaborate repeatedly. But outside of this special case, the cyber world isn't self-policing. It has no common project that everyone is engaged in, no common purpose that is best achieved by co-operation. It is simply a supermarket that offers goods and services of varying quality and price. But whereas we can shop around for the best deals in the real world of supermarkets, desperation and loneliness make us vulnerable to presumptive over-commitment in the world of online dating.

*

So far, I have focused to exclusion on the processes and mechanisms involved in the judgements we make about romantic partners and in other kinds of close relationships. We haven't, so far, asked why or when the curious phenomenon we know as 'falling in love' evolved to become so crucial a part of our psyche. In the next chapter, I finally turn to consider these last two important evolutionary questions.

10

Evolution's Dilemma

Let not women e'er complain
Fickle man is apt to rove!
Look abroad thro' Nature's range,
Nature's mighty law is change.
　　　　'Let Not Woman E'er Complain'

So far, we have established two things. One is that human mating systems are quite variable, at least in so far as they can switch between monogamy and polygamy. The second is that, despite that, all these mating systems seem to be underpinned by the same psychological processes, the phenomenon we refer to generically as 'falling in love'. Although it has been suggested that falling in love is something that happens in other species, it seems to me that we genuinely have something uniquely human to explain here – if only in terms of the intensity of the effect. In this chapter, however, I want to focus on functional questions: why this curious phenomenon might have arisen in the human lineage. Monogamous mating systems (and hence the pairbonding psychology that underpins them) are very rare in mammals: only about 5 per cent of all mammal species are monogamous. Although more common in our own family, the primates, they are still not *that* common: only 15 per cent of primate species are monogamous, and most of those are either small South American cebid and callitrichid monkeys or gibbons (the only monogamous Old World primate). Even if it is true that whatever we see in other species doesn't come close to the intensity of what we find in humans – although certain South American monkeys bear some surprising similarities to us – what other species do can nonetheless sometimes give us valuable insights into the background circumstances that might have led to the evolution of our own behaviour.

Unpacking the evolutionary question

In the grand evolutionary scheme of things, pairbonds are likely to evolve for just four reasons. One is to allow a male to monopolise mating access to a female, so as to ensure that he, and he alone, has paternity. The second is to reduce the risk of offspring being killed by predators. The third is to reduce the risk that the offspring fall prey to infanticidal males. The fourth is to enable the male to contribute to the business of successful rearing. Each of these predicts a different pattern of pairbonded behaviour, and this allows us to gain some insights into the likely evolutionary origins of romantic relationships.

In the first case, for example, the pairbond is really only in the male's interest, since it doesn't matter too much to the female whom she mates with so long as she mates with someone. This being so, we would expect the male to be the one who works hardest at maintaining the relationship, because he has most at stake if the female wanders off and mates with another male. For this reason, it is usually known as the mate guarding hypothesis. We find exactly such behaviour in the klipspringer antelope, one of the most intensely pairbonded species on earth, as we saw in Chapter 3.

The second and third possibilities are closely related,

because they are concerned with reducing the risk that off-spring might be killed. They differ only in who does the killing: a conventional predator versus a member of your own species. Predation is, of course, always a problem, but it is typically adults and, especially, juveniles that bear the brunt of it. In this case, both parties should have a near-equal interest in any protection that the male might have to offer – the female because it is her life that's at stake, and the male because if his female dies at the hand of a predator he will be left without the opportunity to reproduce (given that all the other females in the population will be 'spoken for'). Thus, the predation risk hypothesis favours both parties being equally committed to the relationship. In contrast, 'predation' by members of your own species most commonly affects babies and young infants. Infanticide by males is a perennial risk for monkeys and apes, mainly because of the length of gestation and lactation (itself a consequence of the need to develop their massive brains). In mammals (including humans), females do not come back into reproductive condition so as to be able to conceive another baby until they have weaned the previous one. This is a consequence of a very simple mechanism: the mechanical stimulation of the nipples when the infant sucks triggers a cascade of hormones that shuts down the ovaries and prevents them kicking into a normal menstrual cycle. The system is incredibly sensitive and depends on a specific rate of suckling (one bout every four hours) rather than on the total amount of time the infant actually spends on the nipple. As the infant begins to suckle less, the rate drops below the critical threshold and the mother's menstrual-cycle machinery awakens from its long slumber and kicks into gear. So getting rid of the baby is a sensible strategy from a new male's point of view because it brings the female(s) back into breeding condition and means he can start reproducing much earlier than he would otherwise be able to do.

Infanticidal males are difficult to deal with, because the whole point of their behaviour is to remove the cause of the mother's temporary infertility so that they can sire offspring with her. They target the mother and harass her until her attention is momentarily distracted and the infant can be grabbed. Females are particularly vulnerable to new males who have just joined their group because they are anxious to get breeding as soon as possible while they are still dominant enough to monopolise access to females. Throwing her lot in with one particular male who can act as a bodyguard to defend her and her infant against either predators or other males is a potentially attractive option for a female. In effect, she trades exclusive sexual access for protection. In contrast, there is much less in it for the male, who might be just as well off continuing to behave infanticidally.

This explanation has inevitably become known as the hired gun (or bodyguard) hypothesis. In this case, the female has more to gain by forming a relationship with a male than does the male. For her, the male functions as a bulwark against those who threaten the life of her infant. For the male, it is still a question of whether being exclusively attached to one female yields more sirings over a lifetime than being promiscuous. For this reason, we might expect the female to be the one who is most assiduous about maintaining and servicing the pairbond.

A derivative case of this general explanation arises when females live in large communities that contain many males. In these cases, they may be subjected to persistent harassment by randy males whenever they are in breeding condition. Wild goats and sheep provide the extreme example of this because the males cluster around any female that comes into oestrus and literally fight over her back to mate with her. I spent many years studying feral goats in Scotland and North Wales and never ceased to be amazed at the utter chaos of the annual rut. There were females running every which way trying to escape from the males, and the males clashing horns in ferocious fights with each other so as to be the first in the queue when they finally caught up with a female.

Harassment induces a stress response, which, in the female, triggers the release of both cortisol (the 'stress hormone') and endorphins in large quantities; these in turn block the release of gonadotrophin-releasing hormone in the brain, and, as a result, the ovaries do not receive the chemical signal from the brain that triggers ovulation, so the female is infertile. One solution to this (though never adopted by goats) is for females to form pairbonds with particular males, who then act as a guard to keep other males away. This seems to be the solution adopted by gorillas, though in this case several females may attach themselves to one male, the biggest local thug they can find. The male doesn't particularly seem to care.

The fourth explanation assumes that pairbonds exist to make biparental care possible. This depends on each sex being able to provide parental care as easily as the other, or at least that a division of labour is possible in which one sex provides the childcare and the other provides food to sustain the mother and infant. The first of these occurs commonly among the birds (and is the main reason why around 85 per cent of all bird species are monogamous), but it is realistically possible only because both components of the rearing process are essentially gender-neutral. For birds, the two key steps are sitting on the nest to incubate the eggs and protect them from nest predators, and ferrying food back to the nestlings once they have hatched. For mammals, the equivalent of the incubation phase is done internally in the womb (and so can only be done by the female), while the second involves lactation (so can also only be done by the female). In other words, this option is a non-starter for mammals, whether human or otherwise. The second possibility is somewhat more feasible and is found in a very small number of mammals. The best known example is the dog family (which includes wolves, foxes, coyotes and dholes, as well as our more familiar family pet). This entire family is, uniquely among the mammals, resolutely monogamous, with no exceptions. The reason is that the male is able to feed both the female and the weaned pups by bringing half-digested meat back to the den while the female is preoccupied with her large litter. The male eats the meat as usual after he has made a successful kill, and then carries it back to the den in his stomach; once back home, he regurgitates his stomach contents for the female and pups. Semi-digested meat makes perfect weaning food for the pups. Such behaviour would not be possible for a grazer or leaf-eater: somehow, wilted or half-digested leaves don't have quite the same appeal as nicely tenderised semi-digested meat.

There is one case among primates that provides a genuine example of this kind of division of labour. Uniquely among the primates, the male of the tiny marmoset and tamarin monkeys of South America does all the carrying of the twin infants. The female only has them for about ten minutes or so at a time to provide access to the breast. Even then, it is the male who decides when the infants have had enough time with mum: he will literally rub them off her until they attach themselves to him. In effect, the female is just the milk machine; the male does all the rest, from nappy-changing to bedtime stories.

But the tamarin strategy works because of two features unique to tamarin reproductive biology: the females give birth to twins and are able to come back into reproductive condition straight away after giving birth rather than having to wait until the babies are weaned. The latter means that a second litter is already on the way while the male is busying himself with his newborns, making it worth his while hanging around. This second litter will be born at about the time the first litter is weaned, so the female has the male on the reproductive equivalent of a treadmill. Because the female can twin thanks to the fact that the male absorbs some of the costs of rearing, the male gains more offspring by being monogamous than he would by the less certain strategy of roving in search of ovulating females.

In all the examples of this fourth case, both sexes have a vested interest in maintaining the pairbond: they could not rear offspring successfully without each other's help. So in this case, we might expect both sexes to be equally predisposed to maintaining the relationship. More importantly, and especially for mammals, once the male is committed to investing time and effort into parental care, he has an interest in making sure that the offspring he is helping to rear are his and not someone else's.

Paternity uncertainty is a fact of life for male mammals, and this places them at a considerable evolutionary disadvantage because investing in another male's offspring is genetic altruism. Males who do so simply encourage other males to abandon their offspring and go roving, leaving the costs of rearing to someone else. In fact, this is exactly what tamarin males do: if there is another male in the group willing to act as a helper to the female and carry the infants, the breeding male will abandon his mate as soon as she has conceived and move on to try to find another female he can take over. This works because high female mortality in relatively stressful conditions means that there is a surplus of males and not enough breeding territories. Young unattached males are willing to join another pair and help out until such time as a breeding territory becomes available. Providing males join the territories of male relatives, they have a genetic interest in the offspring they are helping to rear, and the breeding male is content to leave his offspring to them knowing that they will do a decent job precisely because of that. The female herself probably doesn't care *too* much either way: so long as she has someone to do the nappies, that's good enough.

Putting humans in their place

Conventional wisdom has always assumed that humans evolved pairbonding to allow both parents to care for the young. Of course, no one thought for a minute that the males among our ancient predecessors ran around and did all the nappy-changing – after all, modern males only do that kind of thing in very enlightened societies, and, even then, only when they absolutely have to. Rather, the assumption was that the proper explanation lay in the division of labour: as is the case in all traditional societies, the women dealt with the childcare (*very* occasionally assisted by their husbands) and the men went off hunting to bring home the meat. The proper model was thus the dog model. This view has held sway pretty much ever since, under the influence of Darwin's ideas back in the last quarter of the nineteenth century, people started asking questions about the evolution of human marriage systems.

However, questions were raised about the parental care model by a reconsideration of the energetic costs of large brains. Rob Foley and Phyllis Lee calculated the additional energetic costs (relative to chimpanzees) of rearing human offspring with progressively larger brains. As we might anticipate given the fact that our brains are three times the size of chimpanzee brains, the total costs are very considerable. The calculations suggest that these costs could not have been offset either by male provisioning or by a shift to a higher-quality diet such as a meat-based one. Instead, the only way these costs could have been accommodated is by spreading them out over a proportionately longer period of time – in other words, by slowing down the rate of growth and extending the period of parental care (just as we find in humans).

The parental care model received a further setback when Kristen Hawkes suggested that big-game hunting among hunter-gatherers might not have anything at all to do with provisioning. As we saw in Chapter 4, studies of actual hunts by both South American Ache and East African Hadza hunter-gatherers revealed that men hunting large game didn't in fact bring back enough meat for the effort invested to be worthwhile – and, in any case, what they did bring back, they had to share equally among all the families in the camp. Except in the rather special case of the Inuit inhabiting Arctic environments where there isn't much other than meat to eat during the winter, it's the women in hunter-gatherer societies who provide the bulk of the calories through their gathering activities. It isn't as if the women are stuck in camp unable to go foraging for themselves in the way that female canids are. To Hawkes, the hunting-as-provisioning argument just didn't make any sense. Instead, she suggested, what the males are actually doing is advertising their genetic fitness as a mate. It's a form of sexual display, showing off the quality of one's genes by the risks you can afford to take hunting large, dangerous prey in an environment filled with predators.

Hawkes later went on to show in her studies of the Hadza that grandmothers played a far more important role in supporting their daughters' successful reproduction. From this came the idea that menopause might be an adaptive strategy which allowed women to switch the focus of their reproductive effort from their own reproduction to their daughters'. And so the grandmother hypothesis, as it became known, was born. There has since been considerable evidence that even in agricultural societies, maternal grandmothers (but not always *paternal* grandmothers!) have a significant beneficial effect on the number of children that a woman can rear successfully in her lifetime. In any case, most of the exchange arrangements for shared childcare in societies all around the world, traditional and modern, take place between closely related women. In short, men are superfluous to the whole business of reproduction after conception. Child-rearing can be done much better and more efficiently by women co-operating with each other. Urban black communities in the USA provide a nice example of just how well this can work. In those cases where males are here today and gone tomorrow (even if they don't end up being killed in drugs territory wars), the women often form matrilineal multi-generation households that act as a co-operative economic unit: the daughter does the reproducing, mom goes out to work to earn money and granny runs the house. There are no men in the household at all, and none are needed. Perfection and near domestic bliss.

So, if human males were neither providing significant childcare nor bringing home much in the way of bacon – and, in fact, aren't as much use as grandmothers and sisters as childcare help – why should they bother to form pairbonds? Indeed, why should women even bother to attach themselves emotionally to one?

Of the remaining possibilities, we can dismiss the predation argument quite quickly. The very fact that humans practise division of labour in which men and women forage separately in hunter-gatherer societies (one for meat, the other for vegetable foods) makes predation risk an implausible selection pressure. This is not to say that predation risk isn't an issue in traditional societies. Rather, it is simply that whatever the scale of predation risk might be in these cases, it seems that women-only foraging parties deal with it quite adequately. Most primates deal with predation risk by adjusting group size to match, as Susanne Shultz and I have shown in a number of analyses. However, there isn't any requirement that the group should consist of particular kinds of individuals. By ganging together in groups, women can do the job just as well as men. If there is an issue by virtue of the fact that women are typically smaller and less strong than men, they can easily compensate by going around in proportionately larger groups. It's possible that males are useful as protection against predators for females that live alone, as many monogamous antelope and primates do, but human females do not: they universally live in groups and are very reluctant to travel alone any distance. In short, whatever else it might explain, predation risk is not an explanation for pairbonding in humans.

It's also easy to dismiss the mate guarding hypothesis. In mammals, mate guarding tends to occur in one of two circumstances. One is when females choose (or are forced by ecological conditions) to range separately; the other is when they live in large groups and a male needs to prevent rival males from mating with his female around the time of ovulation. The first has been offered as an explanation for monogamy in many small antelope and primates (including many of the smaller South American monkeys as well as the Asian gibbons). However, while the fact that females range alone in their own territories certainly makes monogamy possible, I have been able to show for the marmosets and tamarins as well as the gibbons that males would be as much as five times better off in terms of the number of offspring sired per year if they went roving instead of staying with one female. Of course, weaker males would do less well because they would keep finding themselves excluded from the females whom more powerful males would now be monopolising. But that's just tough. Evolution isn't particularly interested in the wellbeing of the weaker males.

When mate guarding really does occur, it tends to be in the context of large multi-male groups where males risk their female being mated with by any number of other males. Baboons and chimpanzees provide classic examples among the primates, and both of these solve the problem in much the same way: the male locks onto the female and sticks with her so long as she is in oestrus and willing to mate. Their mating consortships rarely last more than a few days, and then the male moves on in search of another fertile female. Neither of these species have permanent pairbonds – although some baboon females do occasionally form close relationships with particular males that can last for a year or two. Basically, we are left with just one option, the hired gun hypothesis driven by the risk of infanticide and harassment.

Hired guns in a poacher's paradise

It's important to remember that humans live in unnaturally large communities by primate standards, even though in traditional hunter-gatherer societies these individuals do not all live in the same place at the same time. Typically, hunter-gatherer communities number around 150 individuals who are distributed among three or four foraging parties (each of thirty to fifty men, women and children) scattered over several hundred square miles of territory. These foraging parties (or overnight camp groups) are somewhat unstable: there is a slow but steady turnover in membership as individual families decide to come and go. A foraging group of around fifty individuals will probably include ten or twelve men of breeding age. For a man, even if the guys you live with are reasonably trustworthy and willing to leave your spouse alone, there remains the risk that other men from the wider community might wander into your camp while you are away. As

Terry Deacon noted in *The Symbolic Species*, some mechanism is needed to prevent these individuals from carrying off your spouse – or, indeed, your spouse suddenly taking a shine to one of them. It's a veritable poacher's paradise out there. So there is, at the very least, a genuine problem of paternity certainty for the men.

However, the problem is perhaps even more serious from the women's side. In the absence of some Deaconesque mechanism, it becomes a sexual free-for-all. The presence of many men in a group – not to mention the risk of others from the wider community and beyond turning up unexpectedly – raises a serious risk of harassment. Sexual harassment is a general problem for females among those species of animals that have promiscuous mating systems. In the chaos of the feral goat rut, the female's day is so disrupted that her ability to feed is reduced by as much as a half. Luckily, oestrus – the period of female sexual receptivity – only lasts a day in goats, so the female can get her day back to normal fairly quickly once the males have lost interest. But in humans, whose sexual receptivity has been disengaged from the menstrual cycle and is more or less continuous, the problem would go on and on.

The great apes illustrate the scale of the problem. Both chimpanzee and orang utan males frequently harass females when they are in oestrus in an attempt to coerce them into sex. And in one sense at least, the strategy works: it turns out that male chimpanzees actually do gain most copulations with the females they harass most. These attacks invariably disrupt the female's foraging, and not infrequently result in injury to her. Females who are subject to repeated attack also have greatly elevated cortisol levels (indicating high stress levels). Since stress is usually associated with increased risk of infertility, females incur a significant disadvantage from being the focus of the males' attention.

Colonially living birds face the same problem, simply because of the sheer numbers of males hanging around. The little bee eater is a member of a flamboyant family of insect-eating birds that grace the plains of Africa and elsewhere. This particular species nests in large colonies in sandbanks along river courses, with each pair having its own nest burrow. Because the females are harassed by gangs of males so much when they leave the burrow to fly to the feeding grounds and when they later arrive back at the burrow, they usually travel with their mates who act as bodyguards. Like most birds, bee eaters are naturally pair-bonded, so the basis of the solution already existed before the problem arose. Bee eaters have simply tweaked one detail of their behaviour to enable the pair to synchronise their feeding forays so that the male can act as hired gun. Female ducks illustrate what can happen when you don't have a hired gun in a free-for-all mating system. Drakes can get so carried away in fighting off rivals that, between them, they can end up quite literally drowning the poor female over whom they are competing.

Lest you suppose that we as a species are too nice to fall prey to such things or that we have somehow avoided the problem, you need look no further than the criminal statistics. Even in our enlightened times, rape and violence lurk only just beneath the surface of civilised society. And the problem is many times worse when social controls are removed, as in times of war or civil unrest. We tend to forget that most of us live in privileged circumstances where rates of violent crime are lower than they have ever been. And that's just for the folks who live in inner-city urban slums. The rest of us in our leafy suburbs face rates of violent crime that are close to negligible. Out there in real life in many parts of Africa, for example, violence is a fact of daily life. And, of course, this has been true historically in Europe as well. The Thirty Years War that ravaged northern Europe between 1618 and 1648 was one of the most destructive ever experienced. Whole regions of Germany were laid waste and social life disrupted while the opposing Catholic and Protestant armies rampaged back and forth across the North German plain. Nor were modern times exempted: the Russian army was given three days' official grace to rape and pillage at will after it had captured Berlin in 1945. Throughout human history, rampaging armies have rarely if ever limited themselves to stealing the property, food and animals of the luckless peasants on whom they preyed. Even today, the descendants of Genghis Khan and his brothers far outnumber anyone else's within the boundaries of the old Mongol empire.* In the bleak view of the seventeenth-century English political philosopher Thomas Hobbes, it is only the imposition of law and order by civil society that prevents us descending into brutishness. The proverbial thin red (or in our case now blue) line that holds back civil disorder and chaos is very thin indeed. Nowhere is this more true

* Around 7 per cent of all living males within the boundaries of the Great Khan's empire have Y chromosomes directly descended from him and his brothers. That's equivalent to around 0.5 per cent of all the males alive today in the world as a whole. Famously (or infamously), his standard strategy on sacking a city was to put all males to the sword and rape the women, and that was irrespective of whether or not the citizens resisted him.

than when there is a shortage of women. Men without access to women are a particular problem, as demonstrated by Philip Starks and Caroline Blackie in an analysis of US rape statistics between 1960 and 1995. They found that on a state-by-state basis the frequency of rape correlates with the frequency of divorce, and argued that this reflects rising levels of sexual frustration as increasing numbers of men are thrown back into the mating market, in a context where divorced men are anxious to remarry but divorced women – once bitten, twice shy – are not. When the number of men searching for partners increases significantly above the number of women available, trouble is inevitable.

This same point is well illustrated in the history of late medieval Portugal. As we saw in Chapter 6, the shift from partible inheritance to primogeniture among the nobility created an excess of impoverished, over-testosteronised younger sons with few prospects of marriage. They became such a public nuisance that eventually even the king was stung into trying to do something about it. The result, it has been argued, was the launch of the Age of Exploration by the Portuguese. Some historians have even argued that the later Crusades were a joint venture between church and state to solve a similar problem elsewhere in Europe during an earlier period.[*]

That hired guns really do provide women with protection is evident from a number of contemporary examples. Among the Ache of eastern Paraguay, for example, a man who replaces a woman's previous 'spouse' after the latter's

---

[*] The shift to primogeniture began some centuries earlier in northern and central Europe than it did in Portugal, mainly because the Iberian peninsula had been under Muslim rule until the early 1400s.

death or disappearance will frequently kill her dependent offspring on the explicit grounds that he is not willing to pay the costs of rearing another man's child. A man's children are safe only so long as he is alive to protect them. A second example is provided by ethnographic studies of tourism which report that lone young women tourists commonly find it convenient to attach themselves to one particular man in a relationship that, in strictly functional terms, trades sex for protection from excessive harassment by others. Lone women in the absence of their kin groups and social networks are vulnerable prey, and especially in cultures where sex is not so freely available for highly testosteronised young men. Margo Wilson and Susan Mesnick analysed a national sample of over twelve thousand Canadian women and found that women who were single or unattached had experienced between two and a half and five times more sexual harassment during the previous twelve months than attached women, even when controlling for age, income, lifestyle and exposure to risk. In fact, the highest differentials were for women aged thirty-five to fifty-four years. When the Vienna ethologists were studying human courtship behaviour in clubs and dance halls, they found that it was necessary for the women researchers to be obviously accompanied by a male confederate; when they were alone, they were bothered so much by men that they were unable to collect data.

Who loves whom?

While the examples in the previous section provide a prima facie case for the hired gun hypothesis, the real test is whether human pairbonds are strictly bidirectional or biased in favour of one sex. Remember that in the first section of this chapter I set up the four possible explanations for pairbonding and pointed out that the four explanations differ in exactly these terms. Mate guarding would be associated with romantically proactive males, the hired gun hypothesis with proactive females, and the predation risk and biparental care hypotheses would have both sexes being equally committed to maintaining the pairbond. The question is: what do human pairbonds actually look like?

There has been an implicit assumption that human pairbonds are mutual rather than one-sided because both parties co-operate in childcare, and both parties experience the process of 'falling in love'. However, it is not entirely obvious that this is so. Although there are surprisingly few studies of this aspect of relationships in humans, folk psychology has always suggested not only that women effectively control which relationships are allowed to blossom and which not, but that they also tend to work harder at building and servicing romantic relationships than men do. Recent work in the psychology of attachment indicates that men, and especially younger men, are more likely to have 'avoidant' or 'dismissive' attachment styles than women. Individuals who fall into these categories tend to agree with statements like: 'I am comfortable without close emotional relationships', 'It is very important to me to feel independent and self-sufficient' and 'I prefer not to depend on others or have others depend on me'. Not surprisingly, they seek less intimacy with partners, and feel less need of close relationships. This is not a peculiarity of our Western culture: a similar tendency for men to be broadly more dismissive–avoidant than women was reported from a large cross-cultural sample of sixty-two different cultural regions. Although there was considerable variation in the level of dismissiveness across cultures, this particular study found that cultural differences (e.g. in terms of gender inequality or gender stereotyping) explained relatively little of the within- or between-sex differences in relationship dismissiveness. In fact, it turns out that most of the differences between cultures were driven by changes in the women's levels of dismissiveness, and this correlated with environmental stress. Where mortality, fertility and AIDS levels were high, women (and to some extent men) became much more dismissive in their attitudes towards the other sex, and the differences between the sexes narrowed. Again, it looks like women are being more choosy and tweaking their behaviour in response to the circumstances they find themselves in: as men become even more dismissive than they usually are, so the women pull up the protective fences further.

The fact that, as we saw in Chapter 4, women are more choosy in their intimate relationships and that, as we saw in Chapter 7, they feel emotional rejection more deeply than men do would seem to provide further evidence that human pairbonds are more female-biased. In the light of all this, I am led inexorably to the conclusion that pairbonds evolved in humans to solve a problem of male harassment and infanticide risk, probably under circumstances where there were very large numbers of rival males around, and that this arose initially through females latching onto individual males as hired guns. This is exactly what seems to have happened in the gorilla, and, as in their case, gives rise to a one-way bond (female to male) and a starlike social network (few or no connections between the females). Once in place, however, it provides an environment in which males can, if circumstances are right, begin to provide paternal care or otherwise invest in the rearing of their offspring. This has not happened in gorillas, but it has happened in the marmosets and tamarins.

In the marmoset/tamarin case, I have been able to show that pairbonding must have evolved before paternal care. Paternal care seems to have evolved in this case precisely because the female was willing and able to raise her game by both reducing the length of her reproductive cycle (the length of time between successive births) and twinning. Primates typically produce a single baby at a time at intervals of a year or more, making promiscuity a worthwhile strategy for males: as we saw earlier, most primate males could defend a large enough territory to include up to five females. However, by twinning and reducing the birth interval to six months, marmoset and tamarin females are able to stack the odds so heavily that it isn't worth a male's while going roving. He can gain just as many offspring by mating monogamously.

In fact, there are quite a few other similarities between humans and the marmosets and tamarins. Human females, too, have managed to dramatically reduce their birth intervals: ours is around three and a half years in natural populations, compared to around five or six in the great apes. Indeed, given our much bigger brains we probably ought to have a birth interval closer to seven or eight years, so the difference is actually much bigger than seems at first sight to be the case: like the marmosets and tamarins, we have halved our birth intervals. Humans also conceal ovulation so that the male is encouraged to mate through the whole of the menstrual cycle, a pattern that is also – uniquely among the South American monkeys – seen in the marmosets and tamarins. And like them, our monogamy is rather flexible. In both cases, breeding males are prone to disappear and set up new relationships elsewhere, and the breeding systems can flip from monogamy into polygamy or even polyandry when circumstances are appropriate.

The difference in the case of the gorilla seems to be the result of the fact that gorilla males are so big. Because the males in a population differ considerably in physical power, and hence their value as a hired gun, the females prefer to attach themselves to the handful of big males rather than distributing themselves more evenly (in which case some females would end up with puny males, and so do worse than others). It's a version of the polygamy threshold again: there is a point in relative male size where it pays to be a second, third or even fourth female of a big male rather than the only female of a small male. Our ancestors were never as sexually dimorphic as the gorillas and orang utans, and so it seems likely that the advantages of outright polygamy never materialised – at least until the development of agriculture, at which point males suddenly had something to offer that females wanted, namely land. It seems that when this happened, less than ten thousand years ago, we came at it with a more unidirectional pairbond and have not yet had time to adjust to multiple relationships. It is perhaps for this reason that polygamous households in contemporary societies are often more like multiple households under one roof. In the classic forms of polygamy that prevail in Africa, each wife has her own hut in the husband's compound, where she cooks, eats, sleeps and raises her children. The husband visits each wife to eat and then sleep on a strict rota (showing preferences would cause unconscionable trouble among the wives). Only in sororal polygyny (where the co-wives are sisters) is it common for them to live together under the same roof. Co-residence is the case in 81 per cent of societies that practise sororal polygyny, but only in 32 per cent of societies that practise non-sororal polygyny. In a word, sisters generally get on better than unrelated individuals. Indeed, it seems that among those Mormon sects that still practise polygyny, sisters married to the same man bicker less than unrelated co-wives.

So for my money, the evidence is all in favour of the hired gun hypothesis, and there are very few crumbs of support for any of the alternatives.

When did pairbonds evolve?

The question of just when human pairbonds evolved remains a live issue, and one that has until recently been pretty much unanswerable. However, in the last few years, new developments have begun to open up this particular mystery. The major issue has been whether it evolved early or late. Were Lucy and her australopithecine cousins pairbonded four million years ago? Or did pairbonds arise only with the appearance of our own species, anatomically modern humans, some two hundred thousand years ago? There has been a long history of attributing monogamy at least back to the origins of our genus, *Homo*, around 1.8 million years ago. If monogamy really did evolve this early, then ironically it suggests (contrary to what has invariably been claimed) that pairbonding probably had nothing to do with the need for biparental care in rearing our unusually large-brained offspring: brain size only really took off with the appearance of archaic humans around five hundred thousand years ago, and so came onto the scene far too late. Archaic humans include the so-called Heidelberg people, *Homo heidelbergensis*, who represent the common ancestor of both ourselves (*Homo sapiens*) and the European Neanderthals (*Homo neanderthalensis*).

Owen Lovejoy has been a vigorous champion of an early emergence of pairbonded monogamy in the human lineage, and in his latest writing on the subject has argued that pairbonding and male parental care had already evolved by 4.4 million years ago when the early hominid *Ardipithecus ramidus* ranged the woodlands of Africa's eastern corner. In his view, two key pieces of anatomical evidence support this claim since both are indicative of monogamy in other monkeys and apes: one is the fact that males are only slightly larger than females and the other is that the males had lost their large canine teeth, which primate males use for fighting (over females). However, it is one thing to claim that one species, or even one population of a species, had a particular pattern of behaviour, and quite another to claim that all members of the lineage did so. *Ardipithecus* might conceivably have been monogamous, but if so it might have been the exception to a more general polygamous rule. With just one species and a handful of specimens, it's rather hard to know. What we do know is that later species of hominids were quite variable in both body size and sexual dimorphism, although all retained the small canines that still characterise modern humans and distinguish us from all the great apes.

My colleagues Emma Nelson and Susanne Shultz found a novel way of determining the mating system of fossil species. In living primates, the second finger is relatively shorter than the fourth finger in polygamous and promiscuously mating species than in monogamous species, especially in males. Indeed, even in modern humans men have shorter second digits compared to women, whose fingers are near-equal in length. This so-called 2D:4D ratio is driven by fetal testosterone, and reflects the testeronised uterine environment in which the infant males of polygamous species develop, preparing them for the battles ahead. Unfortunately for Lovejoy, analyses of the digit lengths of fossils suggest quite uncompromisingly that not only was *Ardipithecus* polygamous, but so were the Neanderthals and most of the early modern fossil humans of our own species. These had digit ratios close to those of the highly polygamous modern Zulus, who lie on the extreme edge of the modern human distribution, and well within the range of the highly polygamous orang utans and chimpanzees. Contemporary modern humans, while still slightly polygamous by these standards (they are definitely not as monogamous as gibbons, the definitively monogamous ape, whose males have near-equal fingers), are much less so than either their immediate ancestors (the Heidelberg humans) or the Neanderthals. Only *Australopithecus* has fingers suggestive of genuine monogamy – although even they lie in the upper end of the modern human distribution, about halfway between the means for modern humans and gibbons, indicating that they were unlikely to have been as habitually monogamous as modern gibbons. So it rather looks like things have switched about quite a bit during the course of human evolution, albeit with a general tendency for polygamy to predominate. If these data tell us anything, they suggest that a trend towards monogamy occurred rather late, and then only in our own lineage.

The difficulty for us is that genetic monogamy and psychological pairbonding aren't necessarily the same thing. As I have pointed out several times, pairbonding occurs as much in polygamous human cultures as it does in those where monogamy is prescribed. The only certain conclusion is that we seem unable to maintain more than one pairbonded relationship at a time. To the extent that men may be happy to have sex with anyone whether or not they are pairbonded, this may make polygamy (one man married to several wives) easier to maintain than polyandry (one wife married to several men) – as seems to be the case, given the extreme rarity of polyandrous societies.

The critical issue in our case must surely have been the size of our communities, and the fact that increasing community size placed ever-growing stresses on females in terms of the risks of infanticide. We can estimate community sizes for fossil populations because there is a very tight relationship between social group size and neocortex size right across the primates. By combining estimates of brain size from individual fossil hominid populations with the equation for this primate-wide relationship, we can get a pretty good idea of how community size changed during the course of our evolutionary history (these are shown in the graphs in my book *The Human Story*). The short answer is that it pottered along at only slightly larger than the community sizes we find in modern chimpanzees (somewhere between fifty and seventy individuals) right through until about five hundred thousand years ago, and then it suddenly took off, rising steeply towards the 150 that we find in modern humans. If chimpanzee females are teetering on the edge of coping with the levels of harassment they incur in communities which typically number fifty individuals (and around ten to twelve breeding males), then the pressures on human females once community size increased significantly beyond, say, a hundred individuals (with twenty to twenty-five breeding males) must have been increasingly intolerable. The need to find a protector must have become overwhelming. Quite where in this sequence the system would have flipped from conventional chimpanzee-like promiscuity into bonded dyads remains uncertain, but the digit ratio data suggest this didn't happen until quite late on, perhaps as late as the appearance of anatomically modern humans some two hundred thousand years ago. Did we have an intermediate step with gorilla-like polygamy? Somehow I doubt it, because our ancestors never developed the level of sexual dimorphism required to create a gorilla-like polygamy threshold.

Envoi

It has been a long and circuitous route to this conclusion, one in which we have explored the psychological and physiological intricacies of relationships as well as their functional consequences. From a strictly biological point of view, relationships exist to facilitate reproduction, to allow individuals to contribute as best they can to the future gene pool of their species. Yet to produce that effect, biology has to work through complex layers of processes at the physiological and the psychological levels. In doing so, it creates the rich tapestry that we experience as our own private lives. Some have claimed that science destroys the magic and the poetry of our experiences, but that is to misunderstand – and in some cases, wilfully misunderstand – our own psychology. Understanding the machinery that creates our experiences does not, and cannot, change those experiences for us, if only because we experience them as emotions, not as bits of the brain lighting up. We will continue to fall in love despite knowing exactly which bits of the brain fire up when we do so. And we will continue to experience the pain of rejection despite knowing precisely how that feeling is produced in our brains. Poets will continue to excite us by the way their subtle wordplay can evoke memories and rouse emotions in ways that we could never manage to do for ourselves.

Some of the facts we have uncovered along the way will have been familiar from everyday life, even if we didn't quite appreciate just what they involved; others will have surprised us. Some, however, will have been disturbing, perhaps even quite challenging for personal beliefs. Unfortunately, science does not guarantee that what it finds out will necessarily satisfy the preconceptions that we have about our world. We enter the arena in the hopes of discovering more about the world we live in, and we have to accept whatever comes our way impartially and with good grace.

What I hope I have been able to do is to give some sense of the extraordinary advances in understanding that have been made at many different levels over the past decade or so. Most of these discoveries have taken place in isolation from each other, so an important aim of this book has been to try to weave these themes into a single coherent story.

In one sense, that has been a challenge in itself simply because we don't yet understand all the fine details of how physiology relates to psychology, and how psychology relates to behaviour. Nonetheless, what I hope I have been able to convey is that below the magic of what we experience and feel in real life lies a complex and fascinating ebb and flow of chemistry. We speak colloquially of the chemistry of love, and it is indeed so – at several different levels.

Acknowledgements

This book owes a great deal to the many graduate students, postdocs and colleagues who have worked with me over the years and whose work I draw on. The book would not have been possible without them. Thank you all.

Bibliography

The epigraphs at the head of each chapter are all from the poems of Robert Burns (1759–1796), whose deep attachment to his wife Jean Armour never seemed to interfere with his romantic propensity to fall in love with every other woman that came his way.

I Now We Are One

Translations of poetry are by M. Rehman (Amir Khusrau Dehlavi), Paul Brians (de Marchaut), Tambimuttu and G. V. Vaidya (Kālidāsa), www.simonhuggins.com/uricon/classic/anon/sumerian_love_poem .htm and heritage-key.com/world/sumerian-love-poem.

Baumeister, R. F. & Bratslavsky, E. (1999). Passion, intimacy, and time: passionate love as a function of change in intimacy. *Personality and Social Psychology Review* 3: 49–67.

Berscheid, E. (1994). Interpersonal relationships. *Annual Review of Psychology* 45: 79–129.

Broad, K. D., Curley, J. P. & Keverne, E. B. (2006). Mother–infant bonding and the evolution of mammalian social relationships. *Philosophical Transactions of the Royal Society, London* 361B: 2199–214.

Collins, W. A., Welsh, D. P. & Furman, W. (2009). Adolescent romantic relationships. *Annual Review of Psychology* 60: 631–52.

Deacon, T. W. (1997) *The Symbolic Species: The Co-Evolution of Language and the Brain*. New York: Norton.

Dunbar, R. I. M., Cornah, L., Daly, F. & Bowyer, K. (2002). Vigilance in humans: a test of alternative hypotheses. *Behaviour* 139: 695–711.

Dunbar, R. I. M. & Dunbar, P. (1980). The pairbond in klipspringer. *Animal Behaviour* 28: 251–63.

Fisher, H. E. (1992). *Anatomy of Love*. New York: Random House.

Bibliography

Fisher, H. E. (2004). *Why We Love*. New York: Holt.

Fisher, H. E., Aron, A. & Brown, L. L. (2006). Romantic love: a mammalian brain system for mate choice. *Philosophical Transactions of the Royal Society, London* 361B: 2173–86.

Hegner, R. E., Elmen, S. T. & Demong, J. (1982). Spatial organisation of the white-fronted bee-eater. *Nature* 298: 264–6.

Jankowiak, W. R. & Fischer, E. F. (1992). A cross-cultural perspective on romantic love. *Ethnology* 31: 149–55.

Kaplan, R. H. and Toshima, M. T. (1990). The functional effects of social relationships on chronic illness and disability. In B. R. Sarason (ed.), *Social Support: An Interactional View*. Wiley: New York.

Krame, S. N. (1963). *The Sumerians: Their History, Culture and Character*. Chicago: University of Chicago Press.

Kummer, H. (1995). *In Quest of the Sacred Baboon: A Scientist's Journey*. Princeton (NJ): Princeton University Press.

Kummer, H., Goetz, W. & Angst, W. (1974). Triadic differentiation: an inhibitory process protecting pair bonds in baboons. *Behaviour* 49: 62–87.

Levin, J. S. (1994). Religion and health: is there an association, is it valid, and is it causal? *Social Science and Medicine* 38: 1475–82.

Lin, Y. (1961). *The Lolo of Llang Shan*. New Haven: HRAF Press.

Lundström, J. N. & Jones-Gotman, M. (2009). Romantic love modulates women's identification of men's body odors. *Hormones and Behavior* 55: 280–4.

Maner, J. K., Rouby, D. A. & Gonzaga, G. C. (2008). Automatic inattention to attractive alternatives: the evolved psychology of relationship maintenance. *Evolution and Human Behavior* 29: 343–9.

Murray, S. L., Griffin, D. W., Derrick, J. L., Harris, B., Aloni, M. & Leder, S. (2011). Tempting fate or inviting happiness? Unrealistic idealization prevents the decline of marital satisfaction. *Psychological Science* 22: 619–26.

Murray, S. L., Holmes, J. G. & Griffin, D. W. (1996). The benefits of positive illusions: idealization and the construction of satisfaction in close relationships. *Journal of Personality and Social Psychology* 70: 79–98.

Shostak, M. (1981). *Nisa: The Life and Words of a !Kung Woman*. New York: Random House.

Bibliography

Sternberg, R. J. (1997). Construct validation of a triangular love scale. *European Journal of Social Psychology* 27: 313–35.

2 Truly, Madly, Deeply

Amico, J. A., Mantella, R. C., Vollmer, R. R. & Li, X. (2004). Anxiety and stress responses in female oxytocin deficient mice. *Journal of Neuroendocrinology* 16: 319–24.

Bales, K. L., Mason, W. A., Catana, C., Cherry, S. R. & Mendoza, S. P. (2007). Neural correlates of pair-bonding in a monogamous primate. *Brain Research* 1184: 245–53.

Belluzzi, J. D. & Stein, L. (1977). Enkephalin may mediate euphoria and drive-reduction reward. *Nature* 266: 556–8.

Bielsky, I. F., Hu, S.-B., Szegda, K. L., Westphal, H. & Young, L. J. (2004). Profound impairment in social recognition and reduction in anxiety-like behavior in vasopressin V1a receptor knockout mice. *Psychoneuroendocrinology* 29: 483–93.

Birk, L. S., Tan, S. A., Fry, W. F., Napier, B. J., Lee, J. W., Hubbard, R. W., Lewis, J. E. & Eby, W. C. (1989). Neuroendocrine and stress hormone changes during mirthful laughter. *American Journal of Medical Science* 298: 390–6.

Boecker, H., Sprenger, T., Spilker, M. E., Henriksen, G., Koppenhoeffer, M., Wagner, K. J., Valet, M., Berthele, A. & Tolle, T. R. (2008). The runners' high: opioidergic mechanisms in the human brain. *Cerebral Cortex* 18: 2523–31.

Broad, K. D., Curley, J. P. & Keverne, E. B. (2006). Mother–infant bonding and the evolution of mammalian social relationships. *Philosophical Transactions of the Royal Society, London* 361B: 2199–214.

Carter, C. S., DeVries, A. C. & Getz, L. L. (1995). Physiological substrates of mammalian monogamy: the prairie vole model. *Neuroscience and Biobehavioral Reviews* 16: 131–44.

Cohen, E., Ejsmond-Frey, R., Knight, N. & Dunbar, R. I. M. (2010). Rowers' high: behavioural synchrony is correlated with elevated pain thresholds. *Biology Letters* 6: 106–8.

Bibliography

Colt, E. W. D., Wardlaw, S. L. & Frantz, A. G. (1981). The effects of running on plasma β-endorphin. *Life Sciences* 28: 1637–40.

Curly, J. P. & Keverne, E. B. (2005). Genes, brains and mammal social bonds. *Trends in Ecology and Evolution* 20: 561–7.

Depue, R. A. & Morrone-Strupinsky, J. V. (2005). A neurobehavioral model of affiliative bonding: implications for conceptualizing a human trait of affiliation. *Behavioral and Brain Sciences* 28: 313–95.

Ditzen, B., Hoppmann, C. & Klumb, P. (2008). Positive couple interactions and daily cortisol: on the stress-protecting role of intimacy. *Psychosomatic Medicine* 70: 883–9.

Domes, G., Heinrichs, M., Gläscher, J., Büchel, C., Braus, D. F. & Herpertz, S. C. (2007b). Oxytocin attenuates amygdala responses to emotional faces regardless of valence. *Biological Psychiatry* 62: 1187–90.

Domes, G., Heinrichs, M., Michel, A., Berger, C. & Herpertz, S. C. (2007a). Oxytocin improves 'mind-reading' in humans. *Biological Psychiatry* 61: 731–3.

Dunbar, R. I. M. (2010). The social role of touch in humans and primates: behavioural function and neurobiological mechanisms. *Neuroscience and Biobehavioral Reviews* 34: 260–8.

Dunbar, R. I. M., Baron, R., Frangou, A., Pearce, E., van Leeuwan, E., Stow, J., Partridge, G., MacDonald, I., Barra, V. & van Vugt, M. (2011). Social laughter is correlated with elevated pain thresholds. *Proceedings of the Royal Society, London.*

Fisher, H. E., Aron, A. & Brown, L. L. (2006). Romantic love: a mammalian brain system for mate choice. *Philosophical Transactions of the Royal Society, London* 361B: 2173–86.

Garver-Apgar, C. E., Gangestad, S. W., Thornhill, R., Miller, R. D. & Olp, J. J. (2006). Major histocompatibility complex alleles, sexual responsivity, and unfaithfulness in romantic couples. *Psychological Science* 17: 830–5.

Gustavson, A. R., Dawson, M. E. & Bonett, D. G. (1987). Androstenol, a putative human pheromone, affects human (*Homo sapiens*) male choice performance. *Journal of Comparative Psychology* 101: 210–12.

Harbach, H., Hell, K., Gramsch, C., Katz, N., Hempelmann, G.

& Teschemacher H. (2000). β-endorphin (1-31) in the plasma of male volunteers undergoing physical exercise. *Psychoneuroendocrinology* 25: 551–62.

Havlíček, J., Saxton, T. K., Roberts, S. C., Jozifkova, E., Lhota, S., Valentova, J. & Flegr, J. (2008). He sees, she smells? Male and female reports of sensory reliance in mate choice and non-mate choice contexts. *Personality and Individual Differences* 45: 565–70.

Hughes, S. M., Harrison, M. A. & Gallup, G. G. (2007). Sex differences in romantic kissing among college students: an evolutionary perspective. *Evolutionary Psychology* 5: 612–31.

Hurst, J. L., Payne, C. E., Nevison, C. M., Marie, A. D., Humphries, R. E., Robertson, D. H. L., Cavaggioni, A. & Beynon, R. J. (2001). Individual recognition in mice mediated by Major Urinary Proteins. *Nature* 414: 631–4.

Insel, T. R. & Shapiro, L. E. (1992). Oxytocin receptor distribution reflects social organisation in monogamous and polygamous voles. *Proceedings of the National Academy of Sciences, USA* 89: 5981–5.

Insel, T. R. & Young, L. J. (2000). Neuropeptides and the evolution of social behavior. *Current Opinion in Neurobiology* 10: 784–9.

Keverne, E. B., Martensz, N. & Tuite, B. (1989). Beta-endorphin concentrations in cerebrospinal fluid of monkeys are influenced by grooming relationships. *Psychoneuroendocrinology* 14: 155–61.

Kikusui, T., Winslo, J. T. & Mori, Y. (2006). Social buffering: relief from stress and anxiety. *Philosophical Transactions of the Royal Society, London* 361B: 2215–28.

Kosfeld, M., Heinrichs, M., Zak, P. J., Fischbacher, U. & Fehr, E. (2005). Oxytocin increases trust in humans. *Nature* 435: 673–6.

Light, K. C., Grewen, K. M. & Amico, J. A. (2005). More frequent partner hugs and higher oxytocin levels are linked to lower blood pressure and heart rate in premenopausal women. *Biological Psychology* 69: 5–21.

Lim, M. M., Wang, Z., Olazabal, D. E., Ren, X., Terwilliger, E. F. & Young, L. F. (2004). Enhanced partner preference in a promiscuous species by manipulating the expression of a single gene. *Nature* 429: 754–7.

Lundström, J. N. & Jones-Gotman, M. (2009). Romantic love modulates women's identification of men's body odors. *Hormones and Behavior* 55: 280–4.

Machin, A. & Dunbar, R. I. M. (2011). The brain opioid theory of social attachment: a review of the evidence. *Behaviour* 148: 985–1025.

Man, J. (2004). *Genghis Khan: Life, Death and Resurrection*. London: Bantam.

Master, S. L., Eisenberger, N. I., Taylor, S. E., Naliboff, B. D., Shirinyan, D. & Lieberman, M. D. (2009). A picture's worth: partner photographs reduce experimentally induced pain. *Psychological Science* 20: 1316–18.

Miller, R. S. (1997). Inattentive and contented: relationship commitment and attention to alternatives. *Journal of Personality and Social Psychology* 73: 758–66.

Nelson, E. E. & Panksepp, J. (1998). Brain structures of infant-mother attachment: contributions of opioids, oxytocin, and norepinephrine. *Neuroscience and Biobehavioral Reviews* 22: 437–52.

Phelan, M. M., McLean, L., Simpson, D. M., Hurst, J. L., Beynon, R. J. & Lian, L. Y. (2010). 1H, 15N and 13C resonance assignment of darcin, a mouse major urinary protein. *Biomolecular NMR Assignments* 4: 239–41.

Porter, R. H. (1999). Olfaction and human kin recognition. *Genetica* 104: 259–63.

Saxton, T. K., Lyndon, A., Little, A. C. & Roberts, S. C. (2008). Evidence that androstadienone, a putative human chemosignal, modulates women's attributions of men's attractiveness. *Hormones and Behavior* 54: 597–601.

Stoddart, D. M. (1990). *The Scented Ape: The Biology and Culture of Human Scent*. Cambridge: Cambridge University Press.

Theodoridou, A., Rowe, A. C., Penton-Voak, I. S. & Rogers, P. J. (2009). Oxytocin and social perception: oxytocin increases perceived facial trustworthiness and attractiveness. *Hormones and Behavior* 56: 128–32.

Uvnäs-Moberg, K. (1998). Oxytocin may mediate the benefits of positive social interaction and emotions. *Psychoneuroendocrinology* 23: 819–35.

Uvnäs-Moberg, K. & Eriksson, M. (1996). Breastfeeding: physiological, endocrine and behavioural adaptations caused by oxytocin and local neurogenic activity in the nipple and the mammary gland. *Acta Psychologica* 85: 525–30.

Uvnäs-Moberg, K., Widström, A. M., Nissen, E. & Björnell, H. (1990). Personality traits in women 4 days post partum and their correlation with plasma levels of oxytocin and prolactin. *Journal of Psychosomatic Obstetrics and Gynecology* 11: 261–73.

Vlahovic, T. A., Roberts, S. B. & Dunbar, R. I. M. Effects of time and laughter on subjective happiness within different modalities of communication. *Journal of Computer-Mediated Communication* (in press).

Wedekind, C., Seebeck, T., Bettens, F. & Paepke, A. J. (1995). MHC-dependent mate preferences in humans. *Proceedings of the Royal Society, London* 260B: 245–9.

Yamakazi, K., Beauchamp, G. K., Curran, M., Bard, J. & Boyse, E. A. (2000). Parent–progeny recognition as a function of MHC odor type identity. *Proceedings of the National Academy of Sciences, USA* 97: 10500–2.

Zillmann, D., Rockwell, S., Schweitzer, K. & Sundar, S. S. (1993). Does humour facilitate coping with physical discomfort? *Motivation and Emotion* 17: 1–21.

3 The Monogamous Brain

Astington, J. W. (1993). *The Child's Discovery of the Mind*. Cambridge (MA): Cambridge University Press.

Bartels, A. & Zeki, S. (2000). The neural basis of romantic love. *NeuroReport* 11: 3829–34.

Bartels, A. & Zeki, S. (2004). The neural correlates of maternal and romantic love. *NeuroImage* 24: 1155–66.

Dunbar, R. I. M. (1998). The social brain hypothesis. *Evolutionary Anthropology* 6: 178–90.

Dunbar, R. I. M. & Shultz, S. (2007). Understanding primate brain

Bibliography

evolution. *Philosophical Transactions of the Royal Society, London* 362B: 649–58.

Finlay, B. L. & Darlington, R. B. (1995). Linked regularities in the development and evolution of mammalian brains. *Science* 268: 1578–84.

Frith, C. D. & Frith, U. (1999). Interacting minds – a biological basis. *Science* 286: 1692–5.

Gallagher, H. L. & Frith, C. D. (2003). Functional imaging of 'theory of mind'. *Trends in Cognitive Science* 7: 77–83.

Kendrick, K. M., Da Costa, A. P., Broad, K. D., Ohkura, S., Guevara, R., Lévy, F. & Keverne, E. B. (1997). Neural control of maternal behaviour and olfactory recognition of offspring. *Brain Research Bulletin* 44: 383–95.

Kinderman, P., Dunbar, R. I. M. & Bentall, R. P. (1998). Theory-of-mind deficits and causal attributions. *British Journal of Psychology* 89: 191–204.

Lewis, P. A., Rezaie, R., Browne, R., Roberts, N. & Dunbar, R. I. M. (2011). Ventromedial prefrontal volume predicts both understanding of others and social network size. *NeuroImage* 57: 1624–9.

Lindenfors, P. (2005). Neocortex evolution in primates: The 'social brain' is for females. *Biology Letters* 1: 407–10.

Miller, G. (2000). *The Mating Mind: How Sexual Choice Shaped the Evolution of Human Nature*. London: Random House.

Pérez-Barbería, J., Shultz, S. & Dunbar, R. I. M. (2007). Evidence for intense coevolution of sociality and brain size in three orders of mammals. *Evolution* 61: 2811–21.

Powell, J., Lewis, P. A., Dunbar, R. I. M., García-Fiñana, M. & Roberts, N. (2010). Orbital prefrontal cortex volume correlates with social cognitive competence. *Neuropsychologia* 48: 3554–62.

Powell, J., Lewis, P. A., Roberts, N., Garcia-Fiñana, M. & Dunbar, R. I. M. (2011). Orbital prefrontal cortex volume predicts social network size: an imaging study of individual differences in humans (submitted).

Ratiu, P. & Talos, I.-F. (2004). The tale of Phineas Gage, digitally remastered. *New England Journal of Medicine* 351: e21.

Reader, S. M. & Laland, K. N. (2002). Social intelligence, innovation, and enhanced brain size in primates. *Proceedings of the National Academy of Sciences, USA* 99: 4436–41.

Saxe, R., Carey, S. & Kanwisher, N. (2004). Understanding other minds: linking developmental psychology and functional neuroimaging. *Annual Review of Psychology* 55: 87–124.

Shultz, S., Bradbury, R., Evans, K., Gregory, R. & Blackburn, T. (2005). Brain size and resource specialisation predict long-term population trends in British birds. *Proceedings of the Royal Society, London* 272B: 2305–11.

Shultz, S. & Dunbar, R. I. M. (2007). The evolution of the social brain: anthropoid primates contrast with other vertebrates. *Proceedings of the Royal Society, London* 274B: 2429–36.

Shultz, S. & Dunbar, R. I. M. (2010). Encephalisation is not a universal macroevolutionary phenomenon in mammals but is associated with sociality. *Proceedings of the National Academy of Sciences, USA* 107: 21582–6.

Shultz, S. & Dunbar, R. I. M. (2010). Social bonds in birds are associated with brain size and contingent on the correlated evolution of life-history and increased parental investment. *Biological Journal of the Linnean Society* 100: 111–23.

Sol, D., Lefebvre, L. & Rodriguez-Teijeiro, D. J. (2005). Brain size, innovative propensity and migratory behaviour in temperate Palaearctic birds. *Proceedings of the Royal Society, London* 272B: 1433–41.

Stiller, J. & Dunbar, R. I. M. (2007). Perspective-taking and memory capacity predict social network size. *Social Networks* 29: 93–104.

Wimmer, H. & Perner, J. (1983). Beliefs about beliefs: representation and constraining function of wrong beliefs in young children's understanding of deception. *Cognition* 13: 103–28.

4 Through a Glass Darkly

Borgerhoff Mulder, M. (1989). Reproductive success in three Kipsigis cohorts. In T. H. Clutton-Brock (ed.), *Reproductive Success*, 419–38. Chicago: University of Chicago Press.

Borgerhoff Mulder, M. (1990). Kipsigis women's preferences for wealthy men: evidence for female choice in mammals? Behavioral Ecology and Sociobiology, 27: 255–64.

Buss, D. M. (1989). Sex differences in human mate preferences. *Behavioral and Brain Sciences* 12: 1–49.

Cashdan, E. (1996). Women's mating strategies. *Evolutionary Anthropology* 5: 134–43.

www.childstats.gov/americaschildren/tables/phy7b.asp

Grammer, K. (1989). Human courtship behaviour: biological basis and cognitive processing. In A. E. Rasa, C. Vogel and E. Voland (eds), *The Sociobiology of Sexual and Reproductive Strategies*, 147–69. London: Chapman and Hall.

Hawkes, K. (1991). Showing off: tests of another hypothesis about men's foraging goals. *Ethology and Sociobiology* 11: 29–54.

Hewlett, B. S. (1988). Sexual selection and paternal investment among Aka Pygmies. In L. Betzig, M. Borgerhoff-Mulder and P. Turke (eds), *Human Reproductive Behaviour: A Darwinian Perspective*, 263–75. Cambridge: Cambridge University Press.

Hill, K. & Kaplan, H. (1988). Trade-offs in male and female reproductive strategies among the Ache. In L. Betzig, M. Borgerhof-Mulder and P. Turke (eds), *Human Reproductive Behaviour: A Darwinian Perspective*, 277–305. Cambridge: Cambridge University Press.

Iredale, W., van Vugt, M. & Dunbar, R. I. M. (2008). Showing off in humans: male generosity as a mating signal. *Evolutionary Psychology* 6: 386–92.

Kelly, S. & Dunbar, R. I. M. (2001) Who dares wins: heroism versus altruism in female mate choice. *Human Nature* 12: 89–105.

Kenrick, D. T. & Keefe, R. C. (1992). Age preferences in mates reflect sex differences in human reproductive strategies. *Behavioral and Brain Sciences* 15: 75–133.

Kruger, D. J. & Nesse, R. M. (2004). Sexual selection and the male: female mortality ratio. *Evolutionary Psychology* 2: 66–85.

Li, N. P., Griskevicius, V., Durante, K. M., Jonason, P. K., Pasisz, D. J. & Aumer, K. (2009). An evolutionary perspective on humor: sexual selection or interest indication? *Personality and Social Psychology Bulletin* 35: 923.

Lycett, J. & Dunbar, R. I. M. (2000). Abortion rates reflect the optimization of parental investment strategies. *Proceedings of the Royal Society, London* 266B: 2355–8.

Orians, G. H. (1969). On the evolution of mating systems in birds and mammals. *American Naturalist* 103: 589–603.

Pawłowski, B., Atwal, R. & Dunbar, R. I. M. (2007). Gender differences in everyday risk-taking. *Evolutionary Psychology* 6: 29–42.

Pawłowski, B. & Dunbar, R. I. M. (1999a). Withholding age as putative deception in mate search tactics. *Evolution and Human Behavior* 20: 53–69.

Pawłowski, B. & Dunbar, R. I. M. (1999b). Impact of market value on human mate choice decisions. *Proceedings of the Royal Society, London* 266B: 281–5.

Pearce, H. E. (1982). *A Sociological Study of Dating Agencies and 'Lonely Hearts' Columns*. PhD thesis, University of London.

Pennebaker, J. W., Dyer, M. A., Caulkins, R. S., Litowitz, D. L., Ackreman, P. L., Anderson, D. B. & McGraw, K. M. (1979). Don't the girls get prettier at closing time: a country and western application to psychology. *Personality and Social Psychology Bulletin* 5: 122–5.

Ronay, R. & von Hippel, W. (2010). The presence of an attractive woman elevates testosterone and physical risk taking in young men. *Social Psychological and Personality Science* 1: 57–64.

Schmitt, D. P., and 118 others (2003). Universal sex differences in the desire for sexual variety: tests from 52 nations, 6 continents, and 13 islands. *Journal of Personality and Social Psychology* 85: 85–104.

Shackelford, T. K., Schmitt, D. P. & Buss, D. M. (2005). Universal dimensions of human mate preferences. *Personality and Individual Differences* 39: 447–58.

Stone, E. A., Shackleford, T. K. & Buss, D. M. (2007). Sex ratio and mate preferences: a cross-cultural investigation. *European Journal of Social Psychology* 37: 288–96.

Voland, E. (1988). Differential infant and child mortality in evolutionary perspective: data from 17th to 19th century Ostfriesland (Germany). In L. Betzig, M. Borgerhoff-Mulder and P. W. Turke (eds), *Human Reproductive Behaviour: A Darwinian Perspective*, 253–62. Cambridge: Cambridge University Press.

Voland, E. (1989). Differential parental investment: some ideas on the contact area of European social history and evolutionary biology.

In V. Standen and R. A. Foley (eds), *Comparative Socioecology: the Behavioural Ecology of Humans and Other Mammals*, 391–403. Oxford: Blackwell.

Voland, E. & Engel, C. (1990). Female choice in humans: a conditional mate selection strategy of the Krummhörn women (Germany 1720–1874). *Ethology* 84: 144–54.

Waynforth, D. & Dunbar, R. I. M. (1995). Conditional mate choice strategies in humans: evidence from 'lonely hearts' advertisements. *Behaviour* 132: 755–79.

5 Saving Face

Anderson, J. L., Crawford, C. B., Nadeau, J. & Lindberg, T. (1992). Was the Duchess of Windsor right? A cross-cultural study of the socioecology of ideals of female body shape. *Ethology and Sociobiology* 13: 197–227.

Arden, R., Gottfredson, L. S., Miller, G. & Pierce, A. (2009). Intelligence and semen quality are positively correlated. *Intelligence* 37: 277–82.

Bates, T. C. (2007). Fluctuating asymmetry and intelligence. *Intelligence* 35: 41–6.

Bereczkei, T., Gyuris, P. & Weisfeld, G. E. (2004). Sexual imprinting in human mate choice. *Proceedings of the Royal Society, London* 271: 1129–34.

Björntorp, P. (1991). Adipose tissue distribution and function. *International Journal of Obesity* 15: 67–81.

Borodulin, K. (2002). Physical activity, fitness, abdominal obesity, and cardiovascular risk factors in Finnish men and women: the national FINRISK 2002 study. PhD thesis, Helsinki University.

Brown, W. M., Cronk, L., Grochow, K., Jacobson, A., Liu, C. K., Popović, Z. & Trivers, R. (2005). Dance reveals symmetry especially in young men. *Nature* 438: 1148–50.

Coetzee, V., Perrett, D. I. & Stephen, I. D. (2009). Facial adiposity: a cue to health? *Perception* 38: 1700–11.

Daly, M. & Wilson, M. (1982). Whom are newborn babies said to resemble? *Ethology and Sociobiology* 3: 69–78.

Bibliography

Frisch, R. (1978). Population, food intake and fertility. *Science* 199: 22–30.

Gangestad, S. W. & Simpson, J. A. (2000). The evolution of human mating: trade-offs and strategic pluralism. *Behavioral and Brain Sciences* 23: 573–644.

Gangestad, S. W. & Thornhill, R. A. (1997). The evolutionary psychology of extrapair sex: the role of fluctuating asymmetry. *Evolution and Human Behavior* 18: 69–88.

Gangestad, S. W., Thornhill, R. & Yeo, R. A. (1994). Facial attractiveness, developmental stability, and fluctuating asymmetry. *Ethology and Sociobiology* 15: 73–85.

Grammer, K., Kruck, K., Juette, A. & Fink, B. (2000). Non-verbal behaviour as courtship signals: the role of control and choice in selecting partners. *Evolution and Human Behaviour* 21: 371–90.

Grammer, K. & Renninger, L. (2004). Disco clothing, female sexual motivation and relationship status: is she dressed to impress? *Journal of Sex Research* 41: 66–74.

Jasienska, G., Ziomkiewicz, A., Ellison, P. T., Lipson, S. F. & Thune, I. (2004). Large breasts and narrow waists indicate high reproductive potential in women. *Proceedings of the Royal Society, London* 271B: 1213–17.

Jones, D. (1995). Sexual selection, physical attractiveness, and facial neoteny. *Current Anthropology* 35: 723–48.

Kalick, S. M., Zebrowitz, L. A., Langlois, J. H. & Johnson, R. M. (1998). Does human facial attractiveness honestly advertise health? Longitudinal data on an evolutionary question. *Psychological Science* 9: 8–13.

Karremans, J. C., Verwijmeren, T., Pronk, T. M. & Reitsma, M. (2009). Interacting with women can impair men's cognitive functioning. *Journal of Experimental Social Psychology* 45: 1041–4.

Kenrick, D., Gutierres, S. & Goldberg, L. (1989). Influence of popular erotica on judgments of strangers and mates. *Journal of Experimental Social Psychology* 25: 159–67.

Kenrick, D., Montello, D., Gutierres, S. & Trost, M. (1993). Effects of physical attractiveness on affect and perceptual judgments: when social comparison overrides social reinforcement. *Personality and Social Psychology Bulletin* 19: 195–9.

Bibliography

Lassek, W. D. & Gaulin, S. J. C. (2008). Waist–hip ratio and cognitive ability: is gluteofemoral fat a privileged store of neurodevelopmental resources? *Evolution and Human Behavior* 29: 26–34.

Little, A. C., Penton-Voak, I. S., Burt, D. M. & Perrett, D. I. (2003). Investigating an imprinting-like phenomenon in humans: partners and opposite-sex parents have similar hair and eye colour. *Evolution and Human Behavior* 24: 43–51.

Manning, J. T., Scutt, D., Whitehouse, G. H. & Leinster, S. J. (1997). Breast asymmetry and phenotypic quality in women. *Evolution and Human Behavior* 18: 1–13.

Mueller, U. & Mazur, A. (2001). Evidence of unconstrained directional selection for male tallness. *Behavioral Ecology and Sociobiology* 50: 302–11.

Mueller, U. & Mazur, A. (1997). Facial dominance in *Homo sapiens* as honest signaling of male quality. *Behavioral Ecology* 8: 569–79.

Mueller, U. & Mazur, A. (1998). Reproductive constraints on dominance competition in male *Homo sapiens*. *Evolution and Human Behavior* 19: 387–96.

Nettle, D. (2002). Height and reproductive success in a cohort of British men. *Human Nature* 13: 473–91.

Nettle, D. (2002). Women's height, reproductive success and the evolution of sexual dimorphism in modern humans. *Proceedings of the Royal Society, London* 269B: 1919–23.

Pawłowski, B. & Dunbar, R. I. M. (2006). Waist-to-hip ratio versus Body Mass Index as predictors of fitness in women. *Human Nature* 16: 50–63.

Pawłowski, B., Dunbar, R. I. M. & Lipowicz, A. (2000). Tall men have more reproductive success. *Nature* 403: 156.

Penton-Voak, I. S., Jones, B. C., Little, A. C., Baker, S., Tiddeman, B., Burt, D. M. & Perrett, D. I. (2001). Symmetry, sexual dimorphism in facial proportions and male facial attractiveness. *Proceedings of the Royal Society, London* 268B: 1617–23.

Perrett, D. I., May, K. A. & Yoshikawa, S. (1994) Facial shape and judgements of female attractiveness. *Nature* 368: 239–42.

Perrett, D. I., Lee, K. J., Penton-Voak, I. S., Rowland, D. R., Yoshikawa, S., Burt, D. M., Henzi, S. P., Castles, D. L. & Akamatsu,

S. (1998). Effects of sexual dimorphism on facial attractiveness. *Nature* 394: 884–7.

Perrett, D. I., Penton-Voak, I. S., Little, A. C., Tiddeman, B. P., Burt, D. M., Schmidt, N., Oxley, R., Kinloch, N. & Barrett, L. (2002). Facial attractiveness judgments reflect learning of parental age characteristics. *Proceedings of the Royal Society, London* 269B: 873–80.

Platek, S. M., Critton, S. R., Burch, R. L., Frederick, D. A., Myers, T. E., Gallup, G. G. (2003). How much paternal resemblance is enough? Sex differences in hypothetical investment decisions but not in the detection of resemblance. *Evolution and Human Behavior* 24: 81–7.

Platek, S. M., Keenan, J. P. & Mohamed, F. B. (2005). Sex differences in the neural correlates of child facial resemblance: an event-related fMRI study. *NeuroImage* 25: 1336–44.

Pollet, T. & Nettle, D. (2008). Taller women do better in a stressed environment: height and reproductive success in rural Guatemalan women. *American Journal of Human Biology* 20: 264–9.

Roberts, S. C. & Little, A. C. (2008). Good genes, complementary genes and human mate choice. *Genetica* 132: 309–21.

Scheib, J. E., Gangestad, S. W. & Thornhill, R. (1999). Facial attractiveness, symmetry, and cues to good genes. *Proceedings of the Royal Society, London* 266B: 1913–17.

Shepherd, J. A. & Strathman, A. J. (1989). Attractiveness and height: the role of stature in dating preference, frequency of dating and perceptions of attractiveness. *Personality and Social Psychology Bulletin* 15: 617–27.

Singh, D. (1993). Adaptive significance of female physical attractiveness: role of waist-to-hip ratio. *Journal of Personality and Social Psychology* 65: 293–307.

Singh, D. (1995). Female health, attractiveness and desirability for relationships: role of breast asymmetry and waist-to-hip ratio. *Ethology and Sociobiology* 16: 465–81.

Singh, D., Dixson, B. J., Jessop, T. S., Morgan, B. & Dixson, A. F. (2010). Cross-cultural consensus for waist–hip ratio and women's attractiveness. *Evolution and Human Behavior* 31: 176–81.

Singh, D. & Luis, S. (1995). Ethnic and gender consensus for the effect of waist-to-hip ratio on judgment of women's attractiveness. *Human Nature* 6: 55–66.

Thornhill, R. & Gangestad, S. W. (1996). The evolution of human sexuality. *Trends in Ecology and Evolution* 11: 98–102.

Thornhill, R. A. & Gangestad, S. W. (1999). The scent of symmetry: a human sex pheromone that signals fitness? *Evolution and Human Behavior* 20: 175–201.

Thornhill, R. A. & Grammer, K. (1999). The body and face of woman: one ornament that signals quality? *Evolution and Human Behavior* 20: 105–20.

Watson, C. D., DeBruine, L. M., Smith, F. G., Jones, B. C., Vukovic, J. & Fraccaro, P. J. (2011). Like father, like self: emotional closeness to father predicts women's preferences for self-resemblance in opposite-sex faces. *Evolution and Human Behaviour* 32: 70–5.

Wetsman, A. & Marlowe, F. (1999). How universal are preferences for female waist-to-hip ratios? Evidence from the Hadza of Tanzania. *Evolution and Human Behaviour* 20: 219–28.

Wingard, D. L., Berkman, L. F. & Brand, R. J. (1982). A multivariate analysis of health-related practices in a nine-year mortality follow-up of the Alameda county study. *American Journal of Epidemiology* 116: 765–75.

Zahavi, A. and Zahavi, A. (1997). *The Handicap Principle: A Missing Part of Darwin's Puzzle*. Oxford: Oxford University Press.

6 By Kith or by Kin

Bateson, M., Nettle, D. & Roberts, G. (2006). Cues of being watched enhance cooperation in a real-world setting. *Biology Letters* 2: 412–14.

Benenson, J. F. & Christakos, A. (2003). The greater fragility of females' versus males' closest same-sex friendships. *Child Development* 74: 1123–9.

Berkman, L. & Glass, T. (2000). Social integration, social networks, social supports and health. In L. Berkman & I. Kawachi (eds), *Social Epidemiology*, 137–73. New York: Oxford University Press.

Boone, J. L. (1986). Parental investment and elite family structure in preindustrial states: a case study of late medieval–early modern Portuguese genealogies. *American Anthropologist* 88: 859–78.

Brooks, R. (1998). The importance of mate copying and cultural inheritance of mating preferences. *Trends in Ecology and Evolution* 13: 45–6.

Christakis, N. A. & Fowler, J. H. (2009). *Connected: The Surprising Power of Our Social Networks and How They Shape Our Lives.* New York: Little, Brown.

Cohen, S. (2005). The Pittsburgh common cold studies: psychosocial predictors of susceptibility to respiratory infectious illness. *International Journal of Behavioral Medicine* 12: 123–31.

Crook, J. H. & Crook, S. J. (1988). Tibetan polyandry: problems of adaptation and fitness. In L. Betzig, M. Borgerhoff-Mulder and P. Turke (eds), *Human Reproductive Behaviour: A Darwinian Perspective*, 97–114. Cambridge: Cambridge University Press.

Curry, O. & Dunbar, R. I. M. (2011a). Altruism in social networks: evidence for a 'kin premium'. *British Journal of Psychology* (revised and resubmitted).

Curry, O. & Dunbar, R. I. M. (2011b). Altruism in networks: the effect of connections. *Biology Letters* Oct 23;7(5):651–3.

Curry, O. & Dunbar, R. I. M. (2011c). Why birds of a feather flock together? The effects of similarity on altruism in a social network. *Social Networks* (submitted).

DeBruine, L. M. (2005). Trustworthy but not lust-worthy: context-specific effects of facial resemblance. *Proceedings of the Royal Society, London* 272B: 919–22.

Ernest-Jones, M., Nettle, D. & Bateson, M. (2010). Effects of eye images on everyday cooperative behaviour: a field experiment. *Evolution and Human Behavior* 32: 172–8.

Flinn, M. V. & England, B. (1995). Childhood stress and family environment. *Current Anthropology* 36: 854–66.

Garver-Apgar, C. E., Gangestad, S. W., Thornhill, R., Miller, R. D. & Olp, J. J. (2006). Major histocompatibility complex alleles, sexual responsivity, and unfaithfulness in romantic couples. *Psychological Science* 17: 830–5.

Helgason, A., Pálsson, S., Guðbjartsson, D. F., Kristjánsson, Þ. & Stefánsson, K. (2008). An association between the kinship and fertility of human couples. *Science* 319: 813–16.

Jedlicka, D. (1980). A test of the psychoanalytic theory of mate selection. *Journal of Social Psychology* 112: 295–9.

Kiecolt-Glazer, J. K., Loving, T. J., Stowell, J. R., Malarkey, W. B., Lemeshow, S., Dickinson, S. L. & Glaser, R. (2005). Hostile marital interactions, proinflammatory cytokine production, and wound healing. *Archives of General Psychiatry* 62: 1377–84.

McPherson, M., Smith-Lovin, L. & Cook, J. M. (2001). Birds of a feather: homophily in social networks. *Annual Review of Sociology* 27: 415–44.

Madsen, E., Tunney, R., Fieldman, G., Plotkin, H., Dunbar, R. I. M., Richardson, J. & McFarland, D. J. (2007). Altruism and kinship: a cross-cultural experimental study. *British Journal of Psychology* 98: 339–59.

Maner, J. K., Rouby, D. A. & Gonzaga, G. C. (2008). Automatic inattention to attractive alternatives: the evolved psychology of relationship maintenance. *Evolution and Human Behavior* 29: 343–9.

Place, S. S., Todd, P. M., Penke, L. & Asendorpf, J. B. (2010). Humans show mate copying after observing real mate choices. *Evolution and Human Behavior* 31: 320–5.

Roberts, S. B. G. & Dunbar, R. I. M. (2011a). Communication in social networks: effects of kinship, network size and emotional closeness. *Personal Relationships* 18: 439–52.

Roberts, S. B. G. & Dunbar, R. I. M. (2011b). The costs of family and friends: an 18-month longitudinal study of relationship maintenance and decay. *Evolution and Human Behavior* 32: 186–97.

Roberts, S., Dunbar, R. I. M., Pollet, T. & Kuppens, T. (2009). Exploring variations in active network size: constraints and ego characteristics. *Social Networks* 31: 138–46.

Silk, J. B. (1990). Which humans adopt adaptively and why does it matter? *Ethology and Sociobiology* 11: 425–6.

Stansfeld, S. A. (2006). Social support and social cohesion. In M. Marmot & R. G. Wilkinson (eds), *Social Determinants of Health*, 148–71. Oxford: Oxford University Press.

Sutcliffe, A. J., Dunbar, R. I. M., Binder, J. & Arrow, H. (2011). Relationships and the social brain: integrating psychological and evolutionary perspectives. *British Journal of Psychology* 103(2): 149–68.

Uller, T. & Johansson, L. C. (2003). Human mate choice and the wedding ring effect: are married men more attractive? *Human Nature* 14: 267–76.

Vigil, J. M. (2007). Asymmetries in the friendship preferences and social styles of men and women. *Human Nature* 18: 143–61.

Voland, E. (1989). Differential parental investment: some ideas on the contact area of European social history and evolutionary biology. In V. Standen and R. A. Foley (eds), *Comparative Socioecology: the Behavioural Ecology of Humans and Other Mammals*, 391–403. Oxford: Blackwell.

Voland, E. & Dunbar, R. I. M. (1995). Resource competition and reproduction: the relationship between economic and parental strategies in the Krummhörn population. *Human Nature* 6: 33–49.

Westermarck, E. A. (1891). *The History of Human Marriage*. New York: Macmillan.

Wolf, A. P. (1995). *Sexual Attraction and Childhood Association: A Chinese Brief for Edward Westermarck*. Stanford: Stanford University Press.

7 A Cheat by Any Other Name

Anderson, K. J. (2006). How well does paternity confidence match actual paternity? Evidence from worldwide nonpaternity rates. *Current Anthropology* 47: 513–20.

Barrett, L., Dunbar, R. I. M. & Lycett, J. E. (2002). *Human Evolutionary Psychology*. Basingstoke: Macmillan/Palgrave. (For Viking berserkers, see p. 262.)

Blair, R. J. R., Jones, L., Clark, F. & Smith, M. (1997). The psychopathic individual: a lack of responsiveness to distress cues? *Psychophysiology* 34: 192–8.

Brown, P. J. (1986). Cultural and genetic adaptations to malaria: Problems of comparison. *Human Ecology* 14: 311–32.

Brown, W. M. & Moore, C. (2003). Fluctuating asymmetry and romantic jealousy. *Ethology and Human Behavior* 24: 113–17.

Brüne, M. (2001). De Clérambault's syndrome (erotomania) in an evolutionary perspective. *Evolution and Human Behavior* 22: 409–15.

Bibliography

Brüne, M. (2003). Erotomanic stalking in evolutionary perspective. *Behavioral Sciences and the Law* 21: 83–8.

Campbell, A. (2002). *A Mind of Her Own: The Evolutionary Psychology of Women*. Oxford: Oxford University Press.

Daly, M. & Wilson, M. (1988). *Homicide*. New York: Aldine de Gruyter.

Daly, M., Wilson, M. & Weghorst, S. (1982). Male sexual jealousy. *Ethology and Sociobiology* 3: 11–27.

DeBruine, L. M. (2005). Trustworthy but not lust-worthy: context-specific effects of facial resemblance. *Proceedings of the Royal Society, London* 272B: 919–22.

de Quervain, D. J.-F., Fischbacher, U., Treyer, V., Schellhammer, M., Schnyder, U., Buck, A. & Fehr, E. (2004). The neural basis of altruistic punishment. *Science* 305: 1254–8.

Dickemann, M. (1979). Female infanticide, reproductive strategies, and social stratification: a preliminary model. In N. A. Chagnon and W. Irons (eds), *Evolutionary Biology and Human Social Behaviour*, 321–67. North Scituate (MA): Duxbury Press.

Dunbar, R. I. M. (1993). On the evolution of alternative reproductive strategies. *Behavioral and Brain Sciences* 16: 291.

Dunbar, R. I. M. (1991). Sociobiological theory and the Cheyenne case. *Current Anthropology* 32: 169–73.

Dunbar, R. I. M., Clark, A. & Hurst, N. L. (1995). Conflict and cooperation among the Vikings: contingent behavioural decisions. *Ethology and Sociobiology* 16: 233–46.

Eisenberger, N. & Lieberman, M. (2004). Why rejection hurts: a common neural alarm system for physical and social pain. *Trends in Cognitive Sciences* 8: 294–300.

Eisenberger, N. I., Lieberman, M. D. & Williams, K. D. (2003). Does rejection hurt: an fMRI study of social exclusion. *Science* 302: 290–2.

Eslinger, P. J. & Damasio, A. R. (1985). Severe disturbance of higher cognition after bilateral frontal lobe ablation: patient EVR. *Neurology* 35: 1731–41.

Gangestad, S. W. & Simpson, J. A. (2000). The evolution of human mating: trade-offs and strategic pluralism. *Behavioral and Brain Sciences* 23: 573–644.

Bibliography

Gangestad, S. W. & Thornhill, R. A. (1997). The evolutionary psychology of extrapair sex: the role of fluctuating asymmetry. *Evolution and Human Behavior* 18: 69–88.

Garver-Apgar, C. E. Gangestad, S. W., Thornhill, R., Miller, R. D. & Olp, J. J. (2006). Major histocompatibility complex alleles, sexual responsivity, and unfaithfulness in romantic couples. *Psychological Science* 17: 830–5.

Gordon, H. L., Baird, A. A. & End, A. (2004). Functional differences among those high and low on a trait measure of psychopathy. *Biological Psychiatry* 56: 516–21.

Hornak, J., Rolls, E. T. & Wade, D (1996). Face and voice expression identification in patients with emotional and behavioural changes following ventral frontal lobe damage. *Neuropsychologia* 34: 247–61.

Kenrick, D. T., Neuberg, S. L., Zierk, K. L. & Krones, J. M. (1994). Evolution and social cognition: contrast effects as a function of sex, dominance, and physical attractiveness. *Personality and Social Psychology Bulletin* 20: 210–17.

Lycett, J. E., Dunbar, R. I. M. & Voland, E. (2000). Longevity and the costs of reproduction in a historical human population. *Proceedings of the Royal Society, London* 267B: 31–5.

Marlowe, F. (2000). Paternal investment and the human mating system. *Behavioural Processes* 51: 45–61.

Meggitt, M. (1962). *Desert People: Study of the Walbiri Aborigines of Central Australia.* Sydney: Angus & Robertson.

Mehu, M., Grammer, K. & Dunbar, R. I. M. (2007). Smiles when sharing. Evolution and *Human Behavior* 6: 415–22.

Mehu, M., Little, A. C. & Dunbar, R. I. M. (2007). Duchenne smiles and the perception of generosity and sociability in faces. *Journal of Evolutionary Psychology* 7: 183–96.

Mueller, U. & Mazur, A. (2001). Evidence of unconstrained directional selection for male tallness. *Behavioral Ecology and Sociobiology* 50: 302–11.

Murray, S. L., Griffin, D. W., Derrick, J. L., Harris, B., Aloni, M. & Leder, S. (2011). Tempting fate or inviting happiness? Unrealistic idealization prevents the decline of marital satisfaction. *Psychological Science* 22: 619–26.

Pérusse, D. (1993). Cultural and reproductive success in industrial societies: Testing the relationship at the proximate and ultimate levels. *Behavioral and Brain Sciences* 16: 267–322.

Seymour, B., Singer, T. & Dolan, R. (2007). The neurobiology of punishment. *Nature Neuroscience Reviews* 8: 300–11.

Simpson, J. A., Gangestad, S. W., Christensen, P. N. & Leck, K. (1999). Fluctuating asymmetry, sociosexuality, and intrasexual competitive tactics. *Journal of Personality and Social Psychology* 76: 159–72.

Singer, T., Seymour, B., O'Doherty, J. P., Stephan, K. E., Dolan, R. J. & Frith, C. D. (2006). Empathic neural responses are modulated by the perceived fairness of others. *Nature* 439: 466–9.

South, S. C., Trent, K. & Shen, Y. (2001). Changing partners: toward a macrostructural-opportunity theory of marital dissolution. *Journal of Marriage and Family* 63: 743–54.

Stillman, T. F., Maner, J. K & Baumeister, R. F. (2010). A thin slice of violence: distinguishing violent from nonviolent sex offenders at a glance. *Evolution and Human Behavior* 31: 298–303.

Theodoridou, A., Rowe, A. C., Penton-Voak, I. S. & Rogers, P. J. (2009). Oxytocin and social perception: Oxytocin increases perceived facial trustworthiness and attractiveness. *Hormones and Behavior* 56: 128–32.

van Wingen, G., Mattern, C., Verkes, R. J., Buitelaar, J. & Fernández, G. (2010). Testosterone reduces amygdala-orbitofrontal cortex coupling. *Psychoneuroendocrinology* 35: 105–13.

van 't Wout, M. & Sanfey, A. G. (2008). Friend or foe: The effect of implicit trustworthiness judgments in social decision-making. *Cognition* 108: 796–803.

Walum, H., Westberg, L., Henningsson, S., Neiderhiser, J. M., Reiss, D., Igl, W., Ganiban, J. M., Spotts, E. M., Pedersen, N. L., Eriksson, E. & Lichtenstein, P. (2008). Genetic variation in the vasopressin receptor 1a gene (AVPR1A) associates with pair-bonding behavior in humans. *Proceedings of the National Academy of Sciences, USA* 105: 14153–6.

Way, B. M., Taylor, S. E. & Eisenberger, N. I. (2009). Variation in the μ-opioid receptor gene (OPRM1) is associated with dispositional and neural sensitivity to social rejection. *Proceedings of the*

National Academy of Sciences, USA 106: 15079–84.

Wilson, M. & Daly, M. (1998). Lethal and nonlethal violence against wives and the evolutionary psychology of male sexual proprietariness. In R. E. Dobash & R. P. Dobash (eds), *Rethinking violence against women*, 199–230. Thousand Oaks, CA: Sage.

Winston, J. S., Strange, B. S., O'Doherty, J. & Dolan, R. J. (2002). Automatic and intentional brain responses during evaluation of trustworthiness of faces. *Nature Neuroscience* 5: 277–83.

Yalman, N. (1963). On the purity of women in the castes of Ceylon and Malabar. *Journal of the Royal Anthropological Institute* 93: 25–38.

Zubieta, J.-K., Dannals, R. F. & Frost, J. J. (1999). Gender and age influences on human brain mu-opioid receptor binding measured by PET. *American Journal of Psychiatry* 156: 842–8.

Zubieta, J.-K., Ketter, T. A., Bueller, J. A., Xu, Y., Kilbourn, M. R., Young, E. A. & Koeppe, R. A. (2003). Regulation of human affective responses by anterior cingulate and limbic μ-opioid neurotransmission. *Archives of General Psychiatry* 60: 1145–53.

Zubieta, J.-K., Smith, Y. R., Bueller, J. A., Xu, Y., Kilbourn, M. R., Jewett, D. M., Meyer, C. R., Koeppe, R. A., Stohler, C. S. (2002). μ-opioid receptor-mediated antinociception differs in men and women. *Journal of Neuroscience* 22: 5100–7.

8 Sleeping with the Devil

Beit-Hallahmi, B. & Argyle, M. (1997). *The Psychology of Religious Behaviour, Belief and Experience*. London: Routledge.

Brüne M. (2001). De Clérambault's syndrome (erotomania) in an evolutionary perspective. *Evolution and Human Behavior* 22: 409–15.

Brüne, M. (2003). Erotomanic stalking in evolutionary perspective. *Behavioral Sciences and the Law* 21: 83–8.

Cohn., N. (1970). *The Pursuit of the Millennium: Revolutionary Millenarians and Mystical Anarchists of the Middle Ages*. Oxford: Oxford University Press.

d'Aquili, E. & Newberg, A. (1999). *The Mystical Mind: Probing the Biology of Religion*. Minneapolis: Fortress Press.

Deedy, D., Law Smith, M., Kent, J., & Dunbar, R. I. M. (2006). Is priesthood an adaptive strategy? Evidence from a historical Irish population. *Human Nature* 17: 393–404.

Kapogiannis, D., Barbey, A., Su, M., Zamboni, G., Krueger, F. & Grafman, J. (2009). Cognitive and neural foundations of religious belief. *Proceedings of the National Academy of Sciences, USA* 106: 4871–81.

Kring, A. M. & Gordon, A. H. (1998). Sex differences in emotion: expression, experience and physiology. *Journal of Personality and Social Psychology* 74: 686–703.

Muncy, R. L. (1973). *Sex and Marriage in Utopian Communities: 19th Century America*. Bloomington: Indiana University Press.

Newberg, A., d'Aquili, E. & Rause, V. (2001). *Why God Won't Go Away*. New York: Ballantine Books.

St Thérèse of Lisieux. (1951). *The Story of a Soul*. London: Burns Oates.

Todd, P. M. & Miller, G. F. (1999). From Pride to Prejudice to Persuasion: satisficing in mate search. In G. Gigerenzer & P. Todd (eds), *Simple Heuristics That Make Us Smart*, 287–308. Oxford: Oxford University Press.

van Vugt, M. & Ahuja, A. (2010). *Selected: Why Some People Lead, Why Others Follow, and Why It Matters*. London: Profile Books.

Wager, T. D., Luan Phan, K., Liberzon, I. & Taylor, S. F. (2003). Valence, gender, and lateralization of functional brain anatomy in emotion: a meta-analysis of findings from neuroimaging. *NeuroImage* 19: 513–31.

Wilson, C., Wilson, D. & Wilson, R. (1992). *Cults and Fanatics*. London: Magpie Books.

9 Love and Betrayal Online

Bargh, J. A. & McKenna, K. Y. A. (2004). The internet and social life. *Annual Review of Psychology* 55: 573–90.

www.dailymail.co.uk/femail/article-1313105/David-Checkley-conned-

30-women-500-000-victims-ask-Why-SO-gullible.html#ixzz1
C3dOeESs

www.dailymail.co.uk/femail/article-1327282/How-middle-class-mother-Tunbridge-Wells-stupid.html#ixzz1C3d5woLq

Deeley, Q., Daly, E., Asuma, R., Surguladze, S., Giampietro, V., Brammer, M., Hallahan, B., Dunbar, R. I. M., Phillips, M. & Murphy, D. (2008). Changes in male brain responses to emotional faces from adolescence to middle age. *NeuroImage* 40: 389–97.

Donath, J. & boyd, d. (2004). Public displays of connection. *BT Technology Journal* 22: 71–82.

Dunbar, R. I. M., Duncan, N. & Nettle, D. (1995). Size and structure of freely forming conversational groups. *Human Nature* 6: 67–78.

www.internet-love-scams.org/

Nie, N. H. (2001). Sociability, interpersonal relations, and the internet: reconciling conflicting findings. *American Behavioral Scientist* 45: 420–35.

Pearce, H. E. (1982). *A Sociological Study of Dating Agencies and 'Lonely Hearts' Columns*. PhD thesis, University of London.

Pollet, T., Roberts, S. B. G. & Dunbar, R. I. M. (2011). Use of social network sites and instant messaging does not lead to increased offline social network size, or to emotionally closer relationships with offline network members. *Cyberpsychology, Behavior and Social Networks* 14: 253–8.

Roberts, S. B. G. & Dunbar, R. I. M. (2011). The costs of family and friends: an 18-month longitudinal study of relationship maintenance and decay. *Evolution and Human Behavior*.

Valenzuela, S., Park, N. & Kee, K. F. (2009). Is there social capital in a social network site?: Facebook use and college students' life satisfaction, trust and participation. *Journal of Computer-Mediated Communication* 14: 875–901.

Valkenburg, P. & Peter, J. (2009). Social consequences of the internet for adolescents: a decade of research. *Current Directions in Psychological Science* 18: 1–5.

Vlahovic, T. A., Roberts, S. B. & Dunbar, R. I. M. Effects of time and laughter on subjective happiness within different modalities of communication. *Journal of Computer-Mediated Communication* (in press).

Bibliography

10 Evolution's Dilemma

Abbott, D. H., Keverne, E. B., Moore, G. F. & Yodyinguad, U. (1986). Social suppression of reproduction in subordinate talapoin monkeys, *Miopithecus talapoin*. In J. Else & P. C. Lee (eds), *Primate Ontogeny*, 329–41. Cambridge: Cambridge University Press.

Aiello, L. C. & Dunbar, R. I. M. (1993). Neocortex size, group size and the evolution of language. *Current Anthropology* 34: 184–93.

Bereczkei, T. (1998). Kinship network, direct childcare, and fertility among Hungarians and Gypsies. *Evolution and Human Behavior* 19: 283–98.

Bowman, G. (1989). Fucking tourists: sexual relations and tourism in Jerusalem's Old City. *Critique of Anthropology* 9: 77–93.

Burton, L. M. (1990). Teenage childbearing as an alternative life-course strategy in multigeneration black families. *Human Nature* 1: 123–43.

Daly, M. & Wilson, S. (1983). *Sex, Evolution and Behavior*. Belmont (CA): Wadsworth.

Davis, J. N. and Daly, M. (1997). Evolutionary theory and the human family. *Quarterly Review of Biology* 72: 407–35.

Del Giudice, M. (2009). Sex, attachment, and the development of reproductive strategies. *Behavioral and Brain Sciences* 32: 1–67.

Dunbar, R. I. M. (2009a). Deacon's dilemma: the problem of pair-bonding in human evolution. In R. I. M. Dunbar, C. Gamble & J. A. G. Gowlett (eds), *Social Brain, Distributed Mind*, 159–79. Oxford: Oxford University Press.

Dunbar, R. I. M. (2004). *The Human Story: A New View of Human Evolution*. London: Faber and Faber.

Dunbar, R. I. M. (2000). Male mating strategies: a modelling approach. In P. Kappeler (ed) *Primate Males*, 259–68. Cambridge: Cambridge University Press.

Dunbar, R. I. M. (1995a). The mating system of Callitrichid primates. I. Conditions for the coevolution of pairbonding and twinning. *Animal Behaviour* 50: 1057–70.

Dunbar, R. I. M. (1995b). The mating system of Callitrichid primates. II. The impact of helpers. *Animal Behaviour* 50: 1071–89.

Dunbar, R. I. M. (2009b). Why only humans have language. In R.

Botha & C. Knight (eds), *The Prehistory of Language*, 12–35. Oxford: Oxford University Press.

Dunbar, R. I. M., Buckland, D. & Miller, D. (1990). Mating strategies of male feral goats: a problem in optimal foraging. *Animal Behaviour* 40: 653–67.

Dunbar, R. I. M. & Dunbar, P. (1980). The pairbond in klipspringer. *Animal Behaviour* 28: 251–63.

Dunbar, R. I. M. & Shultz, S. (2007). Understanding primate brain evolution. *Philosophical Transactions of the Royal Society, London* 362B: 649–58.

Emlen, S. T. & Wrege, P. H. (1986). Forced copulations and intraspecific parasitism: two costs of social living in the white-fronted bee-eater. *Ethology* 71: 2–29.

Finlay, B. L., Darlington, R. B. & Nicastro, N. (2001). Developmental structure in brain evolution. *Behavior and Brain Sciences* 24: 263–308.

Foley, R. A. & Lee, P. C. (1991). Ecology and energetic of encephalization in hominid evolution. *Philosophical Transactions of the Royal Society, London* 334B: 223–32.

Harcourt, A. H. & Greenberg, J. (2001). Do gorilla females join males to avoid infanticide? A quantitative model. *Animal Behaviour* 62: 905–15.

Hawkes, K. (1991). Showing off: Tests of another hypothesis about men's foraging goals. *Ethology and Sociobiology* 11: 29–54.

Hawkes, K., O'Connell, J. F. & Blurton Jones, N. (1989). Hardworking Hadza grandmothers. In V. Standen and R. A. Foley (eds), *Comparative Socioecology: The Behavioural Ecology of Humans and Other Mammals*, 341–66. Oxford: Blackwell.

Hawkes, K., O'Connell, J. F. & Blurton Jones, N. (1997). Hadza women's time allocation, offspring provisioning, and the evolution of long postmenopausal life spans. *Current Anthropology* 38: 551–77.

Hrdy, S. B. (1981). *The Woman That Never Evolved*. Cambridge (MA): Harvard University Press.

Ivey, P. (2000). Cooperative reproduction in Ituri Forest hunter-gatherers: who cares for Efe infants. *Current Anthropology* 41: 856–66.

Kaptijn, R., Thomese, F., van Tilburg, T. G. & Liefbroer, A. C. (2010). How grandparents matter: support for the cooperative breeding hypothesis in a contemporary Dutch population. *Human Nature* 21: 393–405.

Komers, P. E. & Brotherton, P. N. M. (1997). Female space use is the best predictor of monogamy in mammals. *Proceedings of the Royal Society, London* 264B: 1261–70.

Lovejoy, C. O. (2010). Re-examining human origins in the light of *Ardipithecus ramidus*. *Science* 326: 74e1–e8.

McKinney, F., Derrickson, S. R. & Mineau, P. (1983). Forced copulation in waterfowl. *Behaviour* 86: 250–93.

Man, J. (2004). *Genghis Khan: Life, Death and Resurrection*. London: Bantam.

Marlowe, F. (2000). Paternal investment and the human mating system. *Behavioural Processes* 51: 45–61.

Mesnick, S. L. (1997). Sexual alliances: evidence and evolutionary implications. In P. A. Gowaty (ed), *Feminism and Evolutionary Biology*. London: Chapman and Hall.

Møller, A. P. (1985). Mixed reproductive strategy and mate guarding in a semi-colonial passerine, the swallow Hirundo rustica. *Behavioral Ecology and Sociobiology* 17: 401–8.

Nelson, E., Rolian, C., Cashmore, L. & Shultz, S. (2010). Digit ratios predict polygyny in early apes, *Ardipithecus*, Neanderthals and early modern humans but not in *Australopithecus*. *Proceedings of the Royal Society B–Biological Sciences* 278: 1556–63.

Schmitt, D. P. & 127 others (2003). Are men universally more dismissing than women? Gender differences in romantic attachment across 62 cultural regions. *Personal Relationships* 10: 307–31.

Sear, R., Mace, R. and McGregor, I. A. (2000). Maternal grandmothers improve nutritional status and survival of children in rural Gambia. *Proceedings of the Royal Society, London* 267B: 1641–7.

Shultz, S., Noe, R., McGraw, S. & Dunbar, R. I. M. (2004). A community-level evaluation of the impact of prey behavioural and ecological characteristics on predator diet composition. *Proceedings of the Royal Society, London* 271B: 725–32.

Shultz, S. & Dunbar, R. I. M. (2006). Chimpanzee and felid diet composition is influenced by prey brain size. *Biology Letters* 2: 505–8.

Bibliography

Starks, P. T. & Blackie, C. A. (2000). The relationship between serial monogamy and rape in the United States (1960–1995). *Proceedings of the Royal Society, London* 267B: 1259–63.

Thornhill, R. A. & Palmer, C. T. (2000). *A Natural History of Rape: Biological Bases of Sexual Coercion.* Cambridge (MA): MIT Press.

van Schaik, C. P. & Dunbar, R. I. M. (1990). The evolution of monogamy in large primates: a new hypothesis and some critical tests. *Behaviour* 115: 30–62.

Wilson, M. & Mesnick, S. L. (1997). An empirical test of the bodyguard hypothesis. In P. A. Gowaty (ed), *Feminism and Evolutionary Biology.* London: Chapman and Hall.

Zerjal, T., Xue, Y., Bertorelle, G., Wells, R. S., Bao, W., Zhu, S., Qamar, R., Ayub, R., Mohyuddin, A., Fu, S., Li, P., Yuldasheva, N., Ruzibakiev, R., Xu, J., Shu, Q., Du, R., Yang, H., Hurles, M. E., Robinson, E., Gerelsaikhan, T., Dashnyam, B., Mehdi, S. Q. & Tyler-Smith, C. (2003). The genetic legacy of the Mongols. *American Journal of Human Genetics* 72: 717–21.

Zinovieff, S. (1991). Hunters and hunted: *Kamaki* and the ambiguities of sexual predation in a Greek town. In P. Loizos & E. Papataxiarchis (eds), *Contested Identities: Gender and Kinship in Modern Greece,* 203–20. Princeton: Princeton University Press.

Index

Index

Pearce, Helen, 91
Pencovic, Francis, 202
Pennebaker, James, 105
Penton-Voak, Tan, 53
Penton, Ian, 53
Perfect, Mr/Ms, 211, 212–13
perfume, 47, 54, 118
periaqueductal gray, 81
Perrett, David, 123, 126
personal advertisements, 88–90,
    100–4, 218–19
personality theory, 20
PET (positron emission
    tomography), 46
pheromones, 50, 54
physical appearance, 96, 151,
    168
physiology, and psychology, 264
Pio, Padre, 195
Platek, Steve, 128
Poddar, Prosenjit, 210
poetry: love, 5–7, 29–30, 31;
    religious, 197–8
poets, 9, 22, 29, 31, 60, 83
Poland, 94, 117, 120, 121
Pollet, Tom, 230
polyandry, 158, 257, 261
polygamy, 63, 74–5, 173; in
    humans, 99, 176, 236,
    257–8; polygyny, 95, 99, 186,
    190; threshold, 99, 190, 257,
    262; and wealth inequality,
    99
Polynesians, 24, 116, 186
Portugal, 252; nobilty and
    inheritance 155–6
post-industrial societies, kinship,
    134
Powell, Joanne, 78
predation risk, 238, 246–7, 254
predators, 10, 65–6, 108, 223,
    237–8, 239, 241, 246–7
prefrontal cortex, 79, 80–1, 169,
    172, 188, 189

Pride and Prejudice (Austen), 55,
    84
primates: brain size *vs.* size of
    social groups, 62, 74–5;
    division of labour in,
    241–2; grooming and
    endorphins, 41–2; hired gun
    hypothesis, 239, 249–50,
    253, 254, 257; mate
    guarding hypothesis, 237,
    247–8; monogomy/polygamy
    27, 236, 256; pair bonding,
    26–7, 41, 240; predation
    risk, 246; promiscuity in,
    173, 256, 260, 262; roving
    vs. social strategies, 174;
    vasopressin research, 37,
    38; *see also* baboons;
    callitrichid monkeys;
    chimpanzees; gibbons;
    gorillas; humans;
    marmosets; orang utans;
    tamarin monkeys; titi
    monkeys
Prince, Reverend Henry, 201
promiscuity, 63, 65, 68, 71, 239;
    and brain size, 74–5; in
    humans, 175–7, 199–203;
    and oxytocin/vasopressin, 32,
    36–8; in primates, 173, 256,
    260, 262; and sexual
    harassment, 249
prostitutes, 58
Protestants, 251
psyche, 111, 186, 235
psychiatrists, 17, 209
psychological pain, 41, 167
psychology, 9, 20, 23, 30, 84,
    146, 223, 236, 254
psychopaths, 188–9
puberty, 95, 114, 116, 118, 124
Pygmies, Aka, 103

qawwali tradition, 196–7

304